An
Odyssey
in
Time

DISCARDED
From Nashville Public Library

Property of
The Public Library of Nashville and Davidson County
225 Polk Ave., Nashville, Tn. 37203

DALE A. RUSSELL

An Odyssey in Time: The Dinosaurs of North America

PUBLISHED BY
UNIVERSITY OF TORONTO PRESS
IN ASSOCIATION WITH
NATIONAL MUSEUM
OF
NATURAL SCIENCES

© 1989
National Museum
of Natural Sciences, Canada

First paperback edition 1992
Reprinted 1993

ISBN 0-8020-7718-8

ALL RIGHTS RESERVED
The use of any part of this publication,
reproduced, transmitted in any form or by
any means, electronic, mechanical,
photocopying, recording, or otherwise, or
stored in a retrieval system, without prior
consent of the publisher, is an infringement
of copyright law, Chapter C-30, Revised
Statutes of Canada, 1970.

Canadian Cataloguing in Publication Data

Russell, Dale A., 1937 –
An odyssey in time: the dinosaurs of
North America

Co-published by National Museum of
Natural Sciences.
Bibliography: p.
Includes index.
ISBN 0-8020-5815-9 (bound)
ISBN 0-8020-7718-8 (pbk.)

1. Dinosaurs – North America.
2. Dinosaurs – North America – Evolution.
3. Paleoecology,

1. National Museum of Natural
Sciences (Canada). II. Title.
QE862.D5R8 1989   567.9'7
C89-094101-7

PAINTINGS: Eleanor M. Kish

PHOTOGRAPHS: Harry Foster (1984)
except Dale A. Russell: 23, 25, 39, 48, 53,
55, 59, 60, 65, 67, 69 *(top)*, 89, 97, 116,
121, 122, 125 *(bottom)*, 130, 133, 138-9, 157,
158, 172, 173, 177 *(top)*, 187, 189 /
P.J. Currie: 100 / Robert Fillion: 215, 217 /
David M. Jarzen: 127 / Andrew Leitch: 167 /
J.D. Stewart: 123, 125 *(top)* / Courtesy
Geophysics Division, Geological Survey of
Canada: 45, 205 / Courtesy (US) National
Aeronautics and Space Administration:
1, 21, 206-7, 209, 210

MAPS: E.W. Hearn: 24, 26, 37, 49, 57, 66,
98, 103, 112, 124, 147, 189

COPY-EDITING: John Parry

DESIGN: Frank Newfeld

PRINTED IN HONG KONG BY
BOOK ART INC., TORONTO

Contents

Preface

Who Cares about Dinosaurs?

'Dinosaurs? They're extinct! You can't feed dinosaurs to hungry people, and there are more important things for youngsters to learn in order to prepare for life. When I relax after work we talk about the weather or the traffic, but never about dinosaurs.'

Perhaps the names of those of us who could make the foregoing comment our own are legion. Why is it not the same with our children? They immediately grasp the idea that dinosaurs were living creatures and in their minds easily clothe ancient skeletons with flesh. They dream about dinosaurs, draw them, and some who live in the country assemble their own 'dinosaur skeletons' from the bones of cows and horses. Dinosaurs are a touchstone that separates the mentality of children from that of adults.

By these criteria dinosaur specialists are 'abnormal adults.' The bones they study are usually stored in vast rooms, deep within the basements of museums. There, the air is musty and often very warm because of huge pipes overhead that provide heating for galleries and offices on higher levels. The bones are shelved on row upon row of metal racks, which extend up to the ceiling high above, and far down dimly lit corridors. But visitors in other walks of life are seldom repelled by the presence of so much evidence of archaic death. They are usually spellbound and disposed to linger.

Large bones attracted the attention of native peoples in North America. In 1519, Tlaxclalans brought a huge bone, belonging probably to an extinct elephant, to Cortez. Long before the arrival of Europeans, giant bones were seen weathering out in badlands bordering the Rocky Mountains. They obviously did not belong to familiar animals and must have caused their discoverers to wonder what kind of animals had bones like these. A Sioux legend ascribes the bones to huge serpents that were mortally wounded by lightning and had burrowed into the earth to die. But both the Sioux brave of long ago and the modern businessman visiting the bone store-room would agree that dinosaur bones inspire wonder.

This is not to suggest that dinosaurs mean the same thing to everyone. Eastern North Americans tend to look upon dinosaurs as giant and rather odd inhabitants of a prehistoric zoo, while central Africans, on examining restorations of dinosaurs, are impressed with the animals' great strength and the danger they would pose for humans. British citizens are concerned about the care and feeding of dinosaurs, while western North Americans speculate on how one would hunt them. Because the animals resembled dragons, Chinese tend to view dinosaurs with veneration, while for the French dinosaurs were colossal failures because their brains were so small and they became extinct. Dinosaur specialists, who were born in California or Ontario, in Zimbabwe, the United Kingdom, China, or France, are often preoccupied with trying to decide the correct zoological name to give the dinosaur that left the bone behind, never mind what its habits were. But, of course, dinosaur bones *are* beautiful. Just the sight of one makes a paleontologist's heart race!

Dinosaur specialists share all the infirmities of the human condition. Most would subscribe to the proposition that what remains to be known is like an awesome chasm yawning below the tiny ledge of data upon which we stand, or like a boundless ocean of ignorance extending to the limits of a horizon we cannot see. Yet how often has this writer given the impression of speaking with authority on a world that he has never seen and about which he thinks much less often than might be supposed? Various dinosaur specialists have points of view that are not shared by all their colleagues, and these points of view may well be incorrect or incomplete. It is this writer's idiosyncrasy to be impressed by the notion that evolution is not a random process and that the history of life has been affected in a major way by factors in the tremendous and underappreciated extraterrestrial environment. Among the many recent and excellent books on dinosaurs, written from a variety of other points of view, it is a particular pleasure to recommend *Bones for Barnum Brown* by Roland T. Bird, *Men and Dinosaurs* by Edwin H. Colbert, *The Hot-Blooded Dinosaurs* by Adrian Desmond, *The Successful Dragons* by Christopher McGowan, and *The Illustrated Encyclopedia of Dinosaurs* by David Norman (see full citations in Bibliography).

The present compilation represents an attempt to state in a concise manner what one specialist sees in the record of North American dinosaurs. Dinosaurs can be studied on several levels, each of which is valid. One may ask what different kinds of dinosaurs looked like and give names to the various shapes. One may ask what they did and describe different faunas or ecosystems of dinosaurs. One may ask how they changed and sketch the evolution of dinosaur-dominated ecosystems. One may ask how dinosaurs came to be and what became of them and define the limits of the dinosaurian era. One may ask what dinosaurs mean to us and address their place with man in nature. All these levels are considered in the pages that follow.

The answers resulting from the foregoing levels of questioning are not known with equal certainty. A great mass of data exists on the shape of dinosaur bones and their classification. But consider the often-asked question 'Why were dinosaurs so large?' Some suggest that a large body size would have enabled a dinosaur to feed on low-quality food, which may have been available in abundance, and to defend itself better from predators. This is an ecologically reasonable answer, but it is not (now) easy to prove (by examining its bones) that a dinosaur ate low-quality food and that there was much low-quality food in its environment. Others would say that, in fact, the answer is only a surmise. However, the author has used many similar inferences, which seem to him reasonable, in order to attempt to visualize the biology of dinosaurs. A particularly useful source of information lies in the statistically significant correlations between body weight and many behavioural characteristics of living animals. The verbal and visual reconstructions presented here will, it is hoped, stimulate new reconstructions that will be better constrained by newly discovered information.

Although it is far removed from us in time, the dinosaurian world is near to North Americans in space. Dinosaur bones and footprints have been found in many areas of our continent, and they occur often in sediments deposited in ancient environments that were interesting both in geography and in time. An attempt has been made to emphasize geography, so that readers will have a better opportunity to appreciate the history of the dinosaurian era as it pertains to the region in which they live. Some parks and public areas contain a particularly important fragment of the dinosaurian record. The names and thicknesses of geological units in which the dinosaurian record is preserved are also cited in order to engender appreciation for the sedimentary texts in which the history of life on Earth is written. Please do not excavate dinosaur bones. Although mankind did not exterminate the dinosaurs, dinosaur bones are a resource that should be available to everyone. Call your local public authorities should you discover something that may be of importance. Please be aware that most major museums in North America do not have an active program of dinosaurian research, and the staff of those that do are often overtaxed by requests for information. Our museums need the encouragement of their public to initiate or re-establish scholarly centres of dinosaurian expertise.

There are large gaps in the record of North American dinosaurs. In an attempt to fill these gaps, and thereby to provide more continuity, reference has been made to the record of dinosaurs on other continents. However, as valuable as a global perspective surely is, the fascinating dinosaur faunas of Africa, Asia, Europe, and South America are only marginally described. The elucidation of the dinosaurian record as it is preserved in North America is also an ongoing process, and unpublished information – resulting from studies in collaboration with John Horner, on hadrosaur reproduction, and with Robert Long, on the terminal Cretaceous Hell Creek dinosaurs of Montana – has been incorporated into the text. The faunistics of Morrison (Late Jurassic) dinosaurs have not been well studied. Some effort was accordingly made to glean possible associations between the dinosaurs from quarry data (either published or in museum records; see also Coe et al. 1987 in references to chapters 3 and 4). The record of dinosaurs is much richer in late Cretaceous than in older strata, and some localities have been omitted in order to present a regional perspective. Time correlations and paleogeography are always in a continual state of revision. In order to work within something of a consensus in these matters, two volumes in the Cambridge Earth Science Series – Harland et al., *A Geologic Time Scale* (1982), and Owen, *Atlas of Continental Displacement, 200 Million Years to the Present* (1983) – have been tapped to provide a basic framework. Expanding-Earth and fixed-radius paleogeographies are depicted in the latter.

It is not possible to write a volume on dinosaurs that would be simultaneously up to date and exhaustively documented. Relevant and available information is growing faster than anyone can write. There are few good intermediate-level publications on dinosaurs, and for this reason the primary technical literature has been almost the only source for basic information used here. These references are cited in the bibliography, and if greater understanding is desired they should be consulted. The references also include citations of the works of colleagues with points of view different from those of the author. Their views have not otherwise been cited because of a desire to present a concise, non-polemical narrative. Great progress has often been made as a result of dissenting opinions, and for this reason references containing views different from those expressed in the pages that follow are flagged (with an asterisk).

If this volume is attractive, it is because of the artistic abilities of the author's collaborators. He has been allowed to marvel at how the bits of skeleton, landscape, and leaf can be blended together by the medium of oil into ancient excitement by Eleanor M. Kish. Our collaboration in making dinosaur reconstructions has vividly demonstrated to him how talent, truth, and hard work can create something surpassing the sum of the parts. She is his eye upon a vanished world, and to her brush he owes images of landscapes he can never see. The cartography of ancient lands has been ably and discerningly drawn by Edward Hearn. The author led and was led by Ron Seguin through an evolutionary adventure that never was, divining the form of a large-brained dinosaur that might have been if dinosaurs had been able to benefit from an additional 65 million years of natural selection. By the author's side in gales, mud, and fair weather was Harry Foster, capturing on film the barren beauty of badlands where dinosaur bones are miraculously still here to be seen, lying on the surface of the ground.

There are many compensations for being a dinosaur specialist. Foremost among these are knowledgeable and supportive colleagues who have allowed me to impose on their most precious commodity, time. My profound gratitude for their expert counsel is acknowledged with both respect and pleasure.

The first draft of this manuscript was greatly improved through the generous, vigorous, and thorough commentaries of several authorities on dinosaurian evolution. These include Philip J. Currie, Peter Dodson, Robert A. Long, Christopher McGowan, and John S. McIntosh. I esteem their scholarship as much as I value their friendship.

Colleagues in several branches of science have generously made their expertise available to me either personally, by reviewing sections of the manuscript, or by providing guidance in fossil-bearing areas. I wish to express my gratitude to them, as well as to acknowledge their generosity, according to their area of professional interest – in the study of fossil fishes: David Bardack, Don McAllister, and J.D. Stewart; in fossil plants: Peter Crane, David Jarzen, Carol Hotton, and Robert Tschudy; in fossil reptiles: Chris Andress, Donald Baird, William T. Blows, Kenneth Carpenter, Sankar Chatterjee, Dan Chure, Edwin H. Colbert, Philip J. Currie, Peter Dodson, Lance Ericksen, Peter Galton, John Horner, James Jensen, Wann Langston, Jr, Martin Lockley, Robert A. Long, John S. McIntosh, Elizabeth McReynolds, James Madsen, Michael Morales, Michael O'Neill, William Orr, Kevin Padian, and David Weishampel; in geology: David L. Jones, Emlyn H. Koster, and Richard St J. Lambert; in physics: Frank Asaro, Ian Halliday, Eric Jones, Peter Millman, and Richard Muller; and in science writing: Jeff Hecht.

Special thanks are due to David Jarzen, and my colleagues in the Paleobiology Division of the National Museum of Natural Sciences in Ottawa, for many interesting informal discussions relating to the content of the text and the paintings. Mrs Bonnie Livingstone, of the Publications Division of this museum, has been a generous source of support and encouragment.

A man is truly blessed in his family. The manuscript was nurtured in an environment of love and constructive counsel that my wife, Janice, and our three children (now in university) Elizabeth, Maria, and Frank have given me.

I sincerely hope this work will not entirely betray the trust of those who have sought to help me. The responsibility for errors is mine alone.

An
Odyssey
in
Time

I am a part of all that I have met;
Yet all experience is an arch wherethro'
Gleams that untravell'd world, whose margin fades
For ever and for ever when I move.
How dull it is to pause, to make an end,
To rust unburnish'd, not to shine in use!
As tho' to breathe were life. Life piled on life
Were all too little, and of one to me
Little remains; but every hour is saved
From that eternal silence, something more,
A bringer of new things; and vile it were
For some three suns to store and hoard myself,
And this gray spirit yearning in desire
To follow knowledge like a sinking star,
Beyond the utmost bound of human thought.

from

ALFRED, LORD TENNYSON,

'ULYSSES'

Maiasaura

LATE CRETACEOUS,
ABOUT 75 MILLION YEARS AGO

Duck-billed dinosaur (*Maiasaura*) parents
prepare to disgorge food into their northern
Montana nests for the hatchlings to feed on.

[1]

Fabrosaur
(modelled after *Scutellosaurus*)

LATE TRIASSIC, ABOUT 225 MILLION YEARS AGO

A small, archaic ornithischian dinosaur
(modelled after the early Jurassic
Scutellosaurus) drinks beneath an
Araucarioxylon log (see also forest behind),
in Petrified Forest National Park, Arizona.
Note giant scouring rushes (*Neocalamites*),
sparsely crowned cycad-like plant
(bennettitalean), and cowturtles (*Placerias*),
at river's edge.

Before
the Mesozoic

Scales of Time

Long ago, before the beginning of written history, strange animals lived on our planet. The lives of these creatures are separated from our own by enormous spans of time. They lived during different periods, which can be arranged into a natural sequence, or a 'prehistory.' It may seem odd that a well-known dinosaur such as *Triceratops prorsus* could be separated in time from another well-known dinosaur called *Allosaurus fragilis* by a greater number of years than separates *Triceratops prorus* from *Homo sapiens*. It is, however, true. These vast intervals of time must be divided into convenient units so that prehistory can be measured.

What unit should be used? A year is a very understandable period of time. Although it seems longer to a child than to a grandparent, members of both generations can easily recall events of the past year. In science it is often useful to think in terms of factors, or powers, of ten. One year thereby becomes ten to the power zero (one-tenth of ten, or 10^0) years. Ten years is also comprehensible, for we can 'see' changes in our bodies within a decade (one ten, or 10^1 years). Ten times a decade is a century (two tens multiplied, or 10^2 years); changes in the history of a nation become apparent within a century. Ten times a century is a millennium (10^3 years), an interval that is significant in the history of entire civilizations. But we are still far from considering increments of time that will be useful to us in measuring prehistory. Millennia are not long enough.

Periods of a thousand millennia, or a million (10^6) years, are the smallest increments of time used in measuring the history of life on Earth. It is not very helpful to visualize a million years as the approximate equivalent of ten thousand human life-spans. Specialists simply regard it as a geo-temporal unit of *one* (or rarely, but more precisely, as one megayear), without thinking of how it dwarfs their own life-spans. Other time scales are more useful in Earth history. Broad changes in the organic world are usually apparent within 10 million (10^7) years, and most of the divisions of geologic time, which are based on organic change, are of approximately this length. The physical evolution of the Earth is best measured in increments of a thousand million, or a billion (10^9), years. The future evolution of the cosmos is usually discussed in terms of intervals of time large enough to stagger even mathematicians.

Evolving Star Dust

Space and time began some fifteen billion years ago in the titanic explosion of the 'Big Bang.' The subsequent evolution of the physical universe, from the formation of photons to the clumping of galaxies, was determined by conditions prevailing within that explosion during the first tiny fraction of a second of its existence. The structure of living organisms is, in turn, linked intimately to the structure of matter. The initial balance between gravity and the forces within the nucleus of an atom had to be very precisely tuned for planets to form and living things to appear on them. The fact that the balance was achieved is astonishing and suggests that the later emergence of life was not accidental. Neither the physical evolution of the cosmos nor the evolution of living things appears to have been haphazard.

Four thousand six hundred million years ago, in our region of the Milky Way Galaxy, a great cloud of star dust was buffeted by the explosion of a nearby star. It collapsed, and the Sun and planets formed in the condensate. Comets, asteroids, and other dusty materials were swept up in the gravitational field of the nascent Earth. Our planet grew slowly through the accumulation of extraterrestrial rubble, until 3900 million years ago, when the bombardment effectively ended and the crust was stabilized. Oceans already blanketed the planet. Within the next 400 million years several varieties of simple one-celled organisms were probably living in the seas.

Incredibly slowly, living things changed. Simple cells became more complex. Then, 750 million years ago, simple, wormlike creatures composed of many complex cells began to make burrows in the floor of the sea. Approximately 600 million years ago, some of their descendants acquired hard parts that could easily be fossilized. Then they rapidly diversified into a host of different marine creatures. The tempo of evolutionary events increased. The land was colonized by simple plants and millipede-like animals 410 million years ago; amphibians appeared in Quebec 365 million years ago; the oldest known reptiles were preserved in hollow tree stumps in Nova Scotia 305 million years ago. Finally, in soft muds beneath a warm lagoon in Kansas, some small bones were buried. The bones belonged to a long-limbed, lizard-like creature (*Petrolacosaurus*) — a remote ancestor of the dinosaurs. The time was 290 million years before the present. Like us all, it too was made ultimately from star dust.

Previously, all primitive reptiles had fed only on insects and other small animals, but now some began to browse on land plants. But plants were a much more abundant source of food. An age of reptiles dawned on our planet, as populations of herbivorous reptiles expanded during their early 'salad years.' Paradoxically, these reptiles belonged not to the lineage leading to modern reptiles, but rather to another lineage leading to mammals. They are known as therapsids or paramammals. Squat, bulky paramammalian herbivores slowly spread from equatorial regions across the lands, becoming more upright and shortening their tails with the passage of time. They were accompanied by car-nivorous paramammals that took on the appearance of reptilian dogs. Some grew sabre teeth in the front of their jaws. Paramammals were enormously successful and dominated broad lowlands bordering the seas.

Another great branch of reptiles, called sauropsids or parareptiles, were small, active carnivores that rose on their hind legs to run. Their remains are rarely found in sediments of the coastal lowlands. However, they do occur abundantly in sediments deposited in marginal, wetter environments, such as those associated with lakes and rift mountains bordering the sea. As the paramammals diversified on the plains, the parareptiles too gave rise to several major kinds of descendants. The first to appear were lizard-like animals, including the most ancient true lizards. The ancestors of the turtle-bodied, snake-necked fish-eating plesiosaurs invaded the seas soon afterward. Simultaneously, archaic crocodile-like flesh-eaters appeared along rivers and streams. It might have seemed that vigorous paramammals would have always prevented parareptiles from dominating the open, fertile plains, just as modern mammals dominate modern reptiles.

Curiously, it was not to happen this way. Twice in the previous 200 million years, before backboned animals were established firmly on the land, the marine world had been shaken by brief but devastating extinctions. The cause of these extinctions is as yet unknown, but their effects are both obvious and global. Two hundred and forty-eight million years ago a gigantic extinction once again profoundly disrupted life in the seas. Some 90 per cent of all the species of organisms inhabiting the oceans of our planet disappeared in a disaster the magnitude of which has not since been equalled on Earth. The cause of this catastrophe is still being debated. By this time animals were well-established on land, and the paramammals were decimated in the extinctions. The disaster altered the subsequent evolution of life on Earth.

[2]

Massospondylus

EARLY JURASSIC, ABOUT 213 MILLION YEARS AGO

Two small prosauropod dinosaurs
(*Massospondylus*) cross a dune high above a soda
lake, in a rift valley of the proto-Atlantic
rift system, near Nova Scotia. Vegetation
is more abundant on the relatively well-
watered rift mountain crests.

The Triassic

The Era of Middle Life

The great catastrophe of 248 million years ago marked a major watershed in prehistory. It abruptly ended the time known as the 'era of ancient animals,' or the Paleozoic, and ushered the survivors into the 'era of middle animals,' or the Mesozoic. The Mesozoic is often called the Age of Reptiles. It was indeed an age of reptiles, but the period of reptilian ascendancy had actually begun 40 million years before the great catastrophe, with the initial expansion of plant-eating reptiles. It is also known as the Age of Dinosaurs, yet there were no dinosaurs on Earth when the Mesozoic began.

By convention, Mesozoic time is divided into three major periods, beginning with the Triassic, which is the oldest, continuing with the Jurassic, and ending with the Cretaceous. These periods are internationally recognized, and familiarity with their names is essential to understanding a large body of technical and popular writing. The names were founded on, respectively, the threefold division of *Triassic* sediments in Germany, on limestones in the *Jura* Mountains of France, and on white chalks (Latin *creta*), including those in sea cliffs along the English Channel. Triassic time lasted between approximately 248 and 213 million years ago, Jurassic time between 213 and 144 million years ago, and Cretaceous time between 144 and 65 million years ago. Thus the periods are of unequal length, lasting, respectively, about 35, 70, and 80 million years.

Leaving aside, for a moment, the terminological conventions applied to the Mesozoic Era and its subdivisions, it will be useful to remember that our subject is really the 'Era of middle life' on planet Earth. It is composed of two great intervals based on the dominant land animals: one of crocodile-like reptiles (Triassic) and another of dinosaurs (Jurassic and Cretaceous). The latter interval is further subdivided into two sub-intervals: one of cold-blooded giant dinosaurs (Jurassic) and one of smaller, incipiently warm-blooded dinosaurs (Cretaceous).

South of the Equator

During the first 23 million years of the Triassic Period, the record of life on land is best preserved in the Southern Hemisphere, particularly in southern South America and southern Africa. Elsewhere, sediments that were deposited on land during this time are usually either buried or long since destroyed through erosion. The evolutionary events that occurred in the Southern Hemisphere were probably very similar to those that took place simultaneously on what has become the land mass of North America.

In order to trace the series of faunal changes that followed the beginning of Triassic time, a few of the major varieties of animals living then should be identified. Because most of them died out long ago we cannot know them as living creatures. Latinized names do not carry the interesting nuances of 'kitten', 'skunk,' or 'hyena.' Further, a simple translation of their classical roots would be misleading, for the first fossils to be discovered of extinct animals are often fragmentary. The names they originally received may refer to the peculiarities of skeletal parts rather than those of whole animals. Nevertheless, names of some kind must be devised.

Triassic paramammals (therapsids) and parareptiles (sauropsids) are usually classified as reptiles, although both groups belong to a somewhat higher level of anatomical specialization than do living reptiles. There were three major varieties of paramammals. The 'vernacular' names suggested here reflect the mixture of reptilian and mammalian

The major divisions of geological time
and major events during
the dinosaurian era in North America

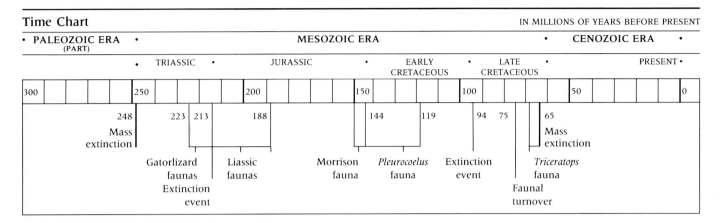

Time Chart									IN MILLIONS OF YEARS BEFORE PRESENT	
• PALEOZOIC ERA • (PART)		MESOZOIC ERA						• CENOZOIC ERA •		
	• TRIASSIC •	JURASSIC		• EARLY CRETACEOUS •	LATE CRETACEOUS •			PRESENT •		

Characteristics they probably exhibited, as well as the appearance of the whole animal. Technical names are given in parentheses.

Cowturtles (dicynodonts) were plant-eating animals generally about the size of small cows. Their bodies were barrel-shaped, and they probably walked with a turtle-like gait. The large heads ended in a toothless beak, although a large upper tusk was often present on each side of the mouth. There was no body armour.

Swinelizards (diademodonts) were in the size range of pigs. They possessed both canine and molar-like teeth within powerful jaws, which were covered by cheeks. Like living hogs they may not have been very discriminating in their choice of food, but they probably subsisted mostly on plants.

Possumlizards (cynognathoids) were small, mammal-like carnivores that later in the Triassic gave rise to true mammals.

In a similar manner, there were three major groups of parareptiles competing with paramammals during Triassic time. The 'vernacular' names are based on a mixture of reptilian or avian characteristics.

Owliguanas (rhynchosaurs) were squat quadrupeds about as large as swinelizards, with barrel-shaped bodies and very broad heads. Their eyes were large and inclined toward the front of the skull. Curved bones that resembled beaver incisors projected forward from the ends of the jaws. The small, peglike teeth were arranged in rows parallel to the edges of the jaws and moved slowly toward the front of the mouth as they were worn down. The powerful construction of the hind limb and the massive claws of the feet suggest that the animals were adapted to 'scratch digging.' They nipped off and chopped up the plant parts on which they fed.

Gatorlizards (thecodonts) took on a variety of body forms but generally resembled long-limbed crocodiles with the foot and general head form of lizards. Their backs were to a greater or lesser extent protected with armour plates and spines. Most were aggressive carnivores, but some fed exclusively on plants.

Dinosaurs are a very well-known group of extinct reptiles. Primitive dinosaurs were small and walked on their hind limbs only. An artificial name such as 'jaylizard' might convey some impression of the appearance of archaic dinosaurs, for most did not resemble very closely their better-known descendants.

As noted before, Triassic time began 248 million years ago in the immediate wake of a gigantic natural disaster. Populations of a small pig-sized cowturtle (*Lystrosaurus*) immediately expanded to fill, albeit somewhat inefficiently, the largely vacant ecological niches for land-dwelling herbivores. Its abundant skeletal remains are found all over the world in terrestrial sediments of this age. Remains of other, larger cowturtles are rare. Several varieties of possumlizards were present, but they were not common. Very few paramammals had survived. Although parareptiles living along the riverbanks and coastlines suffered much less in the extinctions, they were generally small and poorly adapted for bringing down large prey in drier environments. A few heavily built gatorlizards probably frequented the banks of rivers, but no large carnivores lived on the floodplains. There is a single record of a small, primitive owliguana. Large and small flatheaded amphibians dwelled in abundance in streams and lent an aquatic flavour to the entire fauna. Excellent collections of the animal life of this epoch have been made in Antarctica.

Cowturtles soon increased in average bulk but did not increase much in variety. They were joined by great numbers of different kinds of swinelizards, which had descended from smaller possumlizards. Some of the latter in turn became large enough to prey on the big herbivores. Several new kinds of owliguanas appeared. Gatorlizards became more diverse, and some also grew in size.

The swinelizard and cowturtle herds continued in abundance and increased in variety. However, in some environments large owliguanas replaced the other herbivores. All three groups bore massive jaws that were unusually powerful by modern standards. They carried their heads low, and the shortness of their necks indicates that they did not normally feed more than a metre above the ground. The plant structures on which they fed must have been difficult to chew. Judging by the voluminous abdominal cavities of the cowturtles and owliguanas, plant material probably fermented in their bodies before it could be assimilated as food. Some possumlizards grew as large as wolves. Nevertheless, the most impressive land-dwelling carnivores were long-legged, running gatorlizards. Some were small, but others were very large and altogether awesome predators.

Among the many newly evolved terrestrial flesh-eaters were lightly built, agile creatures that preyed on small land animals. These bipedal carnivores (*Staurikosaurus*), measuring a metre and a half in length and weighing 30 kilograms, were the earliest known dinosaurs. Their skeletons were too primitive to be classified within either of the basic groups of dinosaurs, and this fact indicates that all dinosaurs were derived from a common ancestral stock. They would have been fleetingly visible going about their daily pattern of activities some 230 million years before the present.

The first dinosaurs closely resembled small members of several groups of gatorlizards, and it would have been difficult for a casual observer to distinguish them. However, all of them preferred to walk on their hind limbs, and it is through their adaptation to bipedal locomotion that skeletons of primitive dinosaurs can be identified. The backbone was rigidly constructed in order to resist the pull of muscles that connected it to, and propelled, the legs. The pelvis, which served as an anchor for another set of leg muscles, was also solidly attached to at least three vertebral segments of the backbone. Both sets of muscles fastened to the leg bones at positions relatively far down the shaft, increasing their leverage and the efficiency of walking. A boss projected at right angles from the top of the thigh bone to fit into a hole in the centre of the pelvis, and the whole leg swung in a vertical plane about this joint. In most gatorlizards the pelvis formed a solid cap over the end of the thigh-bone. The ankle in dinosaurs was a simple fore-and-aft hinge rather than the more complex structure in gatorlizards. Finally the number of joints in the outside two fingers and toes was

reduced. Any Triassic parareptile skeleton possessing all these attributes is considered to be that of a dinosaur.

Swinelizards and owliguanas continued to be abundant after the appearance of primitive dinosaurs, although cowturtles became less common than formerly. Herbivorous gatorlizards, with their small heads and heavy armour, made their first appearance. Among the carnivores, wolf-sized possumlizards were uncommon, while running gatorlizards were both varied and abundant. The ancestors of true crocodiles also appeared, although their limbs were longer than those of modern crocodilians and the bones of the cheeks and braincase were not so firmly attached to each other.

Some carnivorous dinosaurs (*Herrerasaurus*) grew to lengths of 3 metres and attained weights of 250 kilograms. In these animals the shape of the pelvis generally resembled that of living lizards. They have been termed lizard-hipped, or 'saurischian' dinosaurs. Another major kind of dinosaurs also entered the geological record at about this time, and in these the pelvis was somewhat birdlike. They are called bird-hipped, or 'ornithischian' dinosaurs. The oldest known member of this group (*Pisanosaurus*) was a small herbivorous biped that measured little more than a metre in length. Bird-hipped dinosaurs further differed from lizard-hipped dinosaurs by the greater rigidity of their backbones brought about by the presence of bony tendons. The lower jaws were united in front by a separate beaklike bone, and there were additional bony ossicles attached to the skull above the eyes, in the position of eyebrows.

Reptilian Recovery

The first 23 million years of the Triassic was a time of diversification for terrestrial animal life. A few of the survivors (cowturtles) of the great extinction remained relatively unchanged, but most (possumlizards, gatorlizards) either gave rise to new forms (swinelizards, running gatorlizards, plant-eating gatorlizards, crocodilians, dinosaurs) or acquired the adaptations necessary to enter new environments (owliguanas). This diversification was in response to the disappearance of other animals that formerly had been well established on land. It was not associated with a gradual displacement of archaic creatures by aggressive and highly evolved newcomers. Had there been no previous catastrophe there would have been no subsequent radiation of new forms.

The paramammals that survived the great extinction prospered and increased in size as well as in variety. But

North America,
as seen from *Apollo 16*
in translunar injection

they did not prosper as vigorously as did the parareptilian gatorlizards. Before the great extinction, land faunas were dominated by paramammals, while the trend now seemed to be toward a new age of crocodile-like forms. Dinosaurian bipeds were not very different from their gatorlizard relatives. They remained rare, and fewer than 15 skeletal specimens have so far been collected from sediments deposited during the first 23 million years of Triassic time. Then, as now, most land animals were quadrupeds.

The dominant land plants during this time bore fronds very similar to those of ferns. Unlike ferns, however, they did not propagate themselves by means of microscopic spores that could be spread by the wind. Rather they produced seeds that nourished the developing plantlets during the

critical time when they were becoming established. The plants are therefore known as seed ferns. They are thought for the most part to have taken the form of bushes and small trees, and their fibrous fronds could thus have been browsed by most plant-eating reptiles. Primitive conifers were also present, but these plants grew into trees, partly to carry their foliage out of the reach of large grazing reptiles. Near the end of the first 23 million years of the Triassic the climate was becoming more seasonal or arid in many regions. Seed-fern foliage was becoming coarser still, and these plants were gradually being replaced by taller, cycad-like plants and conifers. A change in the feeding strategy of herbivorous reptiles was indicated.

Meanwhile, the oceans had become populated by a host

of different marine reptiles. Among these were fish-eaters, such as the nothosaurs, with elongated necks, webbed feet, and a body form reminiscent of both otters and lizards. There were also the plesiosaurs, with longer necks and oar-shaped limbs. Armoured and compactly constructed placodonts paddled along reefs, grazing on shellfish which they crushed between their powerful jaws, while dolphin-like ichthyosaurs pursued fish and cephalopods through the open water. In the 23 million years following the great extinction reptiles had established themselves in the seas for the first time and were beginning to flourish again on land.

The record of terrestrial life is well documented in the Northern Hemisphere during the remaining 12 million years of Triassic time.

Birth of a Continent

Friction from tides in the oceans causes the Moon to spin very slowly away from the Earth and the Earth's speed of rotation to decrease very slowly. Because of this, the Moon was closer to the Earth 225 million years ago and would have appeared about one and one-half times larger from the surface of our planet. The Earth was also spinning more rapidly about its axis then, and there were only 22 hours and 45 minutes in each day. Days would have been perceptibly shorter. Modern plants with life cycles regulated by the number of daylight hours would be thrown out of physiological equilibrium if they were transplanted into otherwise favourable Triassic environments. It would be fascinating to know if the radius of the Earth itself was smaller then, and if the planetary atmosphere was denser. Further, an observer in space, looking at the Earth from a position in the plane of its equator, would have seen alternately a hemisphere covered by ocean and a hemisphere covered entirely by land. Some basic attributes of the physical environment were clearly different then.

For perhaps the first time in the history of life on Earth, all the continents were combined into a single gigantic land mass. This supercontinent extended along the equator for 18,000 kilometres, crossing terrain that has since become northern South America, northern Africa, the Arabian Peninsula, and southern China. From the equator, the supercontinent continued toward both the North and South poles, which were located near the same terrains that now make up the arctic and antarctic regions. There had been no ice sheets anywhere on land for 30 million years, nor would there be for more than 200 million years to come. The enormous extent of land on opposite sides of the equator

facilitated the spread of monsoonal climates, where heavy rains followed annual periods of drought. Arid lands were more extensive during late Triassic time. However, there were no great obstacles to prevent land animals from dispersing across the supercontinent. The same varieties of animals inhabited areas that have since become separated by oceans. This is remarkable when one considers how out of place a Nova Scotian moose would seem in Uganda today, or an Argentinian guanaco in Bavaria, or a Cape buffalo in Arizona.

Having just been formed, the great supercontinent was beginning to show the traces of its impending fragmentation. Rift mountains divided broad expanses of land into four major subregions, and would later become the sites of the Atlantic and Indian oceans. At this time, however, the mountain chains were narrow and resembled the mountains and valleys of the great rift zone in the interior of modern east Africa. Rift valleys permitted narrow bodies of marine water to penetrate deeply into the interior of the supercontinent. There, a combination of heat and aridity produced high rates of evaporation. The resulting brines were replenished by water flowing through narrow isthmuses and huge amounts of salt crystallized from them. The arid climates of the time caused salt to be deposited in shallow seas covering the edges of the supercontinent as well. During Triassic time the oceans lost an amount of salt equal to 10 per cent of the salt content of the modern oceans. At no other time during the last 600 million years have the oceans lost nearly so much salt. One series of rift mountains zigzagged across the northwest corner of the supercontinent, bringing brines into the regions where its ends approached the ancient coast. To an observer in space, the trace of this irregular chain would have suggested the eastern margin of modern North America.

The birth of North America began with a strait, comparable in width to the Great Lakes or the Gulf of St Lawrence, which extended south between present-day Greenland and Norway. It ended in bays lined with lime muds and salt north of the future position of Iceland. Rift ridges and valleys fanned out across the plains linking Great Britain and Newfoundland and turned toward the southwest as they entered the tropics to separate the US Atlantic states from the western margin of the Sahara. Great freshwater lakes formed in rift valleys in the southern part of this region. The rift zone then curved around Florida, passing between the Gulf states and the land masses of Cuba and the Yucatan Peninsula of Mexico, both then part of the northern edge of South America. The rift system evidently ended in the central highlands of Mexico in a northwest-

Lake Turkana,
seen from Lapurr Range,
rift mountains, northern Kenya

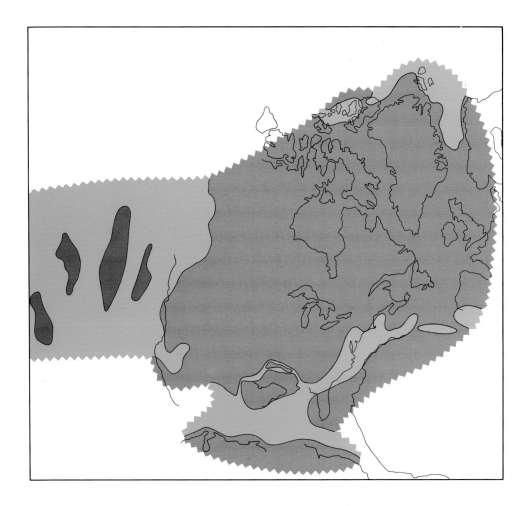

225 million years ago

Break-up of a supercontinent. Seas entering rift valleys begin to separate North America from northern part of supercontinent. Pacific islands may have lain further east than shown.

southeast extending series of transform faults that resembled the San Andreas fault system on the west coast of California today.

The western edge of North America was incomplete. A large river drained plains in the interior of the southwestern United States, passing through Utah on its way to the sea in southern Nevada. Further north, in western Nevada and southwestern Idaho, a range of low but rugged hills bordered the open ocean. Shallow seas met flat coastal lands along the Idaho panhandle and beyond, along the western edge of Alberta. Beaches became sandier in northeastern British Columbia, where the coastline passed near low mountains to the east. The climate was hot enough for salt deposits to form in the region, at that time in the vicinity of 60 degrees north latitude. Rugged coastlines continued on through central Alaska, which then curved to the northeast, instead of the northwest as it does today.

Nearly the entire Rocky Mountain region as we now know it was separated into groups of islands spread 4000 kilometres across the Pacific Ocean to the southwest. Movements of huge plates within the crust of the Earth were creating rifts in the interior of the old supercontinent and bringing its western edge closer to the most easterly of these islands. Beyond the 'horn' of Alaska a shallow sea extended over the polar region. The climate was warm enough to permit lime muds to form beneath its waters, for the last time in Earth history. Rivers as large as the Mississippi flowed from highlands in the rift zones to the east and built huge deltas into the arctic sea in what is now the western Canadian arctic islands. The North Pole lay beyond the sea in eastern Siberia.

Fish-Lizards

Marine reptiles swam in the warm polar seas. These were ichthyosaurs, which were similar in size and shape to modern dolphins. Scattered bones have been collected, and iso-

Sunset over
volcanic islands,
Indonesia

lated skeletons observed on the arctic islands of Canada. Ten individual ichthyosaur skeletons were found fossilized on a bank of hardened lime mud in the Peace River Canyon near Hudson's Hope in northwestern British Columbia. The animals measured from 2 to 10 metres in length and had apparently been stranded together and died.

The enormous 'Rocky Mountain Archipelago' lay close to the tropics and even extended south of the equator. Crushed remains of islands that once measured more than 500 kilometres in diameter have been identified in the Canadian Rockies. Most were mountainous and volcanic, although at least one was flat and resembled the Bahamas. Rains washed nutrients from the volcanic islands into the surrounding seas. These in turn teemed with life, including swarms of free-floating shelled relatives of octopi and squids. Ichthyosaurs flourished among the islands as well.

In the Wallowa Mountains of northeastern Oregon, small, 2-metre-long ichthyosaurs swam in a shallow sea surrounding one of these volcanic island groups. Skeletons

the size of small whales have been collected in northern California, and gigantic specimens 15 metres long occur in Berlin-Ichthyosaur State Park in central Nevada. The latter animals also apparently died as a result of many separate strandings in shallow water. Bony plates that strengthened the walls of the eye indicate that the animals possessed enormous eyes – 20 centimetres in diameter. Twigs and logs of archaic conifers were found near the site where the ichthyosaurs died. Salt crystals on the bones indicate that the climate was hot and dry. Although the California and Nevadan ichthyosaur sites are within about 500 kilometres of each other today, they were probably much further apart and surely on different islands in the Pacific 225 million years ago.

One of the most desolate fossil localities in the world is situated on a small island in a frigid sea about 80 kilometres west by northwest of the North Magnetic Pole, which is to say 80 kilometres west by northwest of nowhere. Here, in 1853, members of a British fleet in search of a lost expedition discovered a single bone from the neck of a land-dwelling reptile of late Triassic age. It was named *Arctosaurus* and was for many years believed to be the first dinosaur bone collected in Canada. In fact, it belongs to a trilophosaur, a lizard-like herbivore with kangaroo-like teeth (the grinding surfaces of the teeth in kangaroos and in trilophosaurs are formed from transverse ridges). From this point to the south for thousands of kilometres no other remains of terrestrial animals of this age have ever been collected.

In western Colorado, near the town of Gateway, red muds deposited in quiet waters about 230 million years ago (Moenkopi Formation, 30 metres thick) are crossed with fossil trackways. Some of these were made by gatorlizards. Others are tridactylate prints made by erect, bipedal creatures. They constitute the oldest evidence of dinosaurs in North America, occurring in one of two regions near the southern limits of the then-forming continent where fossils of land-dwelling organisms may be found. One region is the southwestern United States and adjacent Mexico; and the other lay along the old rift valleys of Atlantic Canada and the Atlantic seaboard of the United States.

Wolf-Crocodiles

During late Triassic time a huge topographic basin was centred on what is now the Colorado Plateau, then near the northern edge of the tropical zone. It measured 1250 kilometres across along the shore of the western sea and extended 1500 kilometres to the southeast over Utah, Wyoming, Colorado, Arizona, and New Mexico to western Texas. To the south was a range of volcanic mountains that paralleled the border between the United States and Mexico. To the southeast were the eroded roots of ancient mountains. To the north was a vast expanse of low, rolling plains. The centre of the basin was interrupted by a range of mountains in central Colorado and western New Mexico.

The late Triassic record began here with sediments deposited around a large lake in Wyoming (Popo Agie Formation, 75 metres thick) and with an alternating sequence of stream and lake sediments in Texas and New Mexico (Dockum Formation, 500 metres thick). In both areas the fauna was dominated by gatorlizards, and the rapid evolution of these reptiles can be used to document the passing

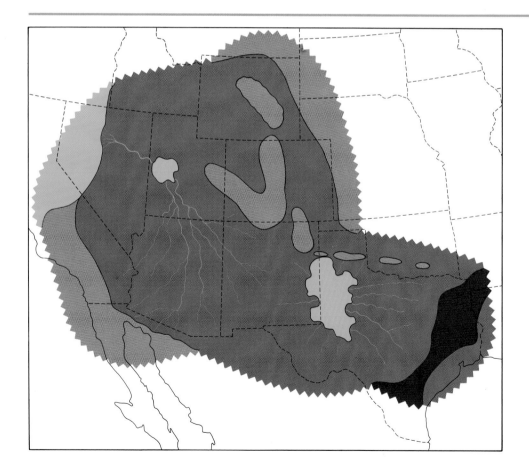

220 million years ago

A vast topographic basin centres on what is now US southwest.

Fabrosaur:
detail of animal drinking from a stream,
Petrified Forest National Park, Arizona, 225 million
years ago, during late Triassic time

of time. Bones of large cowturtles have been found in Wyoming. The more spectacular remains there include those large carnivorous gatorlizards (rauisuchians) that resembled a cross between a crocodile and a wolf. Pieces of the skull of these creatures can be mistaken for those of carnivorous dinosaurs, for the animals ranged between 2.5 and 6 metres in length.

In western Texas and eastern New Mexico, a large lake filled and dried with regional changes in rainfall. The most abundant fossils preserved in its sediments belonged to many different kinds of fish, including freshwater sharks, primitive heavily scaled fishes, lobe-finned fishes, and lungfish. Skeletons of 3-metre-long trilophosaurs document a distribution that extended from the arctic nearly to the equator. The lake shore was evidently frequented by owlizards as well. In the southern part of the Texas panhandle, 60 kilometres southeast of Lubbock, a remarkable site has yielded the remains of terrestrial animals that were overwhelmed in a flash flood. Bones have been identified of lizard-like reptiles, small possumlizards, gatorlizards, and a few belonging to a small saurischian and a 2-metre-long ornithischian dinosaur (*Technosaurus*).

Skeletal parts of a large rauisuchian (*Postosuchus*) are present, indicating an animal 4 metres long and weighing more than 200 kilograms. With them are remains of small,

at least partly bipedal gatorlizards, which resembled small carnivorous dinosaurs and are known elsewhere in greater completeness. Gatorlizard relationships in these creatures are clearly indicated by the structure of the ankle, the five-fingered hand, and the presence of bony scutes along the back.

The most spectacular bones are, however, those of two crow-sized animals. These specimens are still in the process of being carefully studied, but they exhibit many characteristics that still occur in one large group of living vertebrates. Among these are the presence of large eye sockets, a wishbone, and a keeled breastbone. The skull bears teeth, the fingers were clawed, and the tail was very long. It would appear that the animal resembled ancestral birds at least as closely as did the famous *Archaeopteryx*, which was preserved in sediments deposited about 75 million years after the burial of the Texas specimens.

A Petrified Forest

As the fossil record temporarily came to an end in Texas and Wyoming, it began in regions further to the west. Vari-coloured clays from weathered volcanic ash rapidly spread into western regions of the basin from the highlands to the southwest. They accumulated in thicknesses of up to 300 metres (Petrified Forest Member, Chinle Formation). As the southern borderlands were slowly eroded away, the basin continued to fill gradually with red alluvium (red beds part, Chinle Formation). These silts were frequently inter-layered with carbonate ledges formed in broad, shallow lakes. The red sediments also reached a thickness of 300 metres near the southern slopes of the mountains in Col-orado. The fossils these strata contain change as the vari-coloured clays merge upward into the red beds.

The varicoloured clays are now exposed in beautiful badlands across northern Arizona and New Mexico, and into southern Utah. They have yielded abundant evidence of the plants and animals that once populated the basin. Shells of freshwater clams, lungfish teeth, and globs of fossil excrement containing bits of bone are everywhere. The lat-ter were excreted in quiet water by aquatic fish-eating pre-dators. Amphibians measuring 3 metres in length (*Metoposaurus*) frequented sheltered margins of streams. There they were able to camouflage themselves from prey which they snapped up in their huge, flat heads. When these peripheral bodies of water dried up, as they did from time to time, the skeletons of dozens of these creatures were left clumped together in the last part of the pond to dry out. Their feeble limbs were inadequate to carry them more than a short distance over the ground. Six metre-long gatorlizards resembling gavials (phytosaurs, *Rutiodon*), sculling with compressed tails, actively pursued their prey through more open water. They carried their nostrils above the sur-face of the water far behind the tips of their jaws, in a position just in front of their eyes. Giant specimens could attain lengths of 8 to 9 metres.

On the lands bordering the rivers grew strange plants (bennettitaleans) with thick stems and crowns of small fronds reminiscent of those of palms. The most ancient squat stems of true cycads (*Charmorgia*) grew among them. The areas between the heavy-stemmed plants were blanketed by sev-eral different varieties of ferns. Dense stands of giant scour-ing rushes (*Neocalamites*) spread across abandoned sand-bars by means of subterranean runners. Their stems were as much as 40 centimetres in diameter and 8 metres tall. Clus-ters of tree ferns filled moist hollows along the stream banks. Rooting and browsing paths through the low vegetation were heavily armoured gatorlizards (aetosaurs, *Calyptosu-chus, Desmatosuchus, Paratypothorax*). Their heads were small and, in side view, showed some similarity to those of stur-geon; from above, the widened cropping beak resembled the muzzle of a cow. Their 4-metre-long bodies were thick-ened in midsection to accommodate the more massive

Large trunk, in Petrified Forest
National Park, Arizona:
Long Logs site, Petrified Forest Member,
Chinle Formation, west of Holbrook

Petrified Forest National Park:
Petrified Forest Member, Chinle Formation,
Blue Mesa area. Skeletal remains of large
amphibians (*Metoposaurus*) and gatorlizards
(*Rutiodon*) are abundant here.

digestive system characteristic of a herbivore. Their differing patterns of body armour were fully justified by the presence of rapacious, 6-metre-long wolf-crocodiles. The tooth marks of the carnivores occur on amphibian bones in Germany, where at least one specimen was in turn severely bitten by its prey.

The most famous environment preserved within the varicoloured clays has given Petrified Forest National Park in Arizona its name. In many areas of the park hundreds of fossilized logs measuring up to 140 centimetres in diameter and 35 metres in length are strewn densely across the badlands. Clusters of stumps were buried upright, with some penetrating as much as 3 metres of overlying sediment – in evidence of the rapidity with which sediments accumulated. The forests grew in stable lowlands, far enough away from streams that the trees would usually mature without being toppled by undercutting currents, but close enough so that an adequate supply of ground water was available throughout the year. Frequently, however, meandering streams did cut into mature forests, and created spectacular fields of fossilized logs in the park. The thin cuticles of fossil leaves indicate that the climate was generally humid, but soda deposits in ponds show that periods of seasonal aridity could occur. The abundance of charred wood in the varicoloured clays further suggests that the forests were swept by fires during dry periods.

The forests were spectacular and gothic in appearance, with trees attaining heights of 45 to 60 metres. Groups of stumps preserved in the position in which they grew indicate that large trees were only 3 to 4 metres apart. High overhead the branches must have met, forming a closed canopy and allowing only shafts of light to penetrate to the forest floor. The forests were nearly pure stands of one kind of tree (*Araucarioxylon*) that may have somewhat resembled the South American araucarias but were probably not closely related to them. Only a few per cent of the logs belonged to smaller species (*Woodworthia*, with logs 15 metres long, and *Schilderia*, with logs 9 metres long). If the trees escaped forest fires and undercutting by streams, growth rings indicate they could have lived for hundreds of years. Old trees were probably blown over in storms, possibly after their trunks were weakened by wood-rotting fungi. Fossilized fungal brackets are frequently found on fallen logs. Other trees were killed by insect larvae, the burrows of which girdled the trees beneath the bark. When scattered giants fell, ferns grew in the light beneath the rents in the canopy. The height of the forest trees protected their foliage from browsing reptiles, and their shade inhibited the growth of lower plants on which the reptiles fed. There were probably

Blue Mesa area.
TOP: Detail of Petrified Forest Member (p. 30).
BOTTOM: Light-coloured Sonsela Sandstone,
overlying Petrified Forest Member

Tree trunks, broken and
fossilized, at Long Logs site,
Petrified Forest National Park

few large animals living on the ancient forest floor.

Near the southern edge of the great Arizona basin, lowland forests and streams came into contact with low foothills, which were gradually buried as the basin filled with sediment. Near the buried hills, the sediments contain indications of plants preferring more elevated environments. Their pollen was carried by wind and water toward the basin. These minute grains suggest that the hillsides may have supported many different kinds of primitive cone- and seed-bearing trees. Some bore fronds resembling those of cedars and ginkgos, while other, spirelike trees bore leaves shaped like straps. The well-drained hillsides probably supported more open, shrubby growth. The branches may have been within the reach of large herbivores. Isolated bones of a large cowturtle (*Placerias*) are not common in the varicoloured clays. When they do occur, they are found in the lower levels of the clays near the buried hills. Perhaps the animals browsed in the scrub forests of the low hills.

A large quarry has been excavated into sediments that once formed the floor of a seasonal pond near St Johns, in eastern Arizona. It produced bones from at least 39 individuals of this cowturtle. About half of the animals, presumably males, bore large horns on their upper jaws, which projected forward and down. There are few remains of juveniles. The pond, formerly filled with plant material, had dried out, leaving behind a crust of calcium sulphate on the soil. The malodorous waters of the dwindling pool somehow attracted the herbivores. Their skeletons were trampled, and most of the bones were carried away by scavengers. Those that remain often bear tooth marks. Scattered bones of wolf-crocodiles suggest the identity of some of the scavengers, which evidently also killed each other at the carrion hole.

Small carnivores also visited the drying pond, including a quadrupedal gatorlizard slightly more than a metre long (*Hesperosuchus*). A few bones and teeth of dinosaurs were preserved as well, and they too were derived from small animals. Among them are parts of long-legged archaic carnivorous dinosaurs (staurikosaurs), and two partial skeletons of small saurischians called prosauropods. Their bodies were stouter than those of the carnivore, and the animals were evidently able to nourish themselves on nutritious plant shoots above the reach of the cowturtles. There were a few scattered teeth of tiny herbivorous ornithischians as well. All these dinosaurs were probably agile creatures, much more independent of aquatic environments than were most of their gatorlizard contemporaries.

Toward the northern end of Petrified Forest National Park, the varicoloured clays grade upward into red silts.

Finally, from Chinde Point, the landscape opens up over Lithodendron Wash onto the vast expanse of the Painted Desert. A different assemblage of animals characterized the time of change from varicoloured to red sedimentation, and fossils become less abundant. The most common animal preserved in these strata is a large herbivorous gatorlizard (*Typothorax*), with an armoured, disc-shaped body lacking horns along its perimeter. Remains of gavial-like gatorlizards (*Rutiodon*) are also relatively abundant. Recently, a skeleton of an archaic, long-legged staurikosaur was collected from the badlands below Chinde Point.

The varicoloured clays, with their stream-bank fauna of large amphibians and gatorlizards, are found also in northern New Mexico. Here, upper levels of the red clays merge horizontally into reddish-brown siltstones derived from mountains in Colorado to the north. In these reddish siltstones at Ghost Ranch Quarry Natural Landmark, north of Albuquerque, was found the only known site where dinosaurs occur abundantly in the Triassic of North America.

Complete skeletons representing more than 100 individuals of small carnivorous saurischians were buried together in sediments that accumulated in the wake of a flash flood. They belonged to an animal named *Coelophysis*, which measured 2.5 metres in length and weighed about 70 kilograms. No other North American dinosaur of Triassic age is known from material even remotely as complete as this.

Coelophysis was a graceful creature, although its legs seem rather short in comparison to the length of its body, relative to these proportions in later carnivorous dinosaurs. The fossil remains are obviously unbalanced ecologically, for all the animals were carnivores. Indeed the skeleton of one small dinosaur was found within the body cavity of an adult of the same species. Preservational circumstances suggest that it was in the process of being digested. In spite of the implicit cannibalism, it is difficult to avoid the conclusion that the animals herded together, at least at times, like mongooses. It is also apparent that dinosaurs could be abundant in some areas within a landscape dominated by gatorlizards. Another site of about this age, where a group of about 70 individuals of a dinosaurian herbivore (*Plateosaurus*, a prosauropod) was overwhelmed and buried in a mudflow, occurs in West Germany. At a second site near Ghost Ranch the remains of a 6-metre-long carnivorous dinosaur with an elongated neck (*Longosaurus*) have been collected. Other localities in the vicinity have produced carnivorous (*Rutiodon*) and herbivorous (*Typothorax*) gatorlizard material.

Red silts continued to accumulate over the varicoloured clays throughout most of the great basin. Toward the east,

The Triassic

Painted Desert badlands, eroded into red strata of Chinle Formation, Petrified Forest National Park

Ghost Ranch, near Abuquiu, New Mexico.
LEFT: From bottom: red shales of Chinle Formation; Entrada Sandstone,
a fossil sand sea; shales of Morrison Formation; and Dakota Sandstone.
RIGHT: Ghost Ranch Quarry has yielded many excellent skeletons of *Coelophysis*.

Red strata, Chinle Formation, exposed
at Lacey Point, where skeleton of small
theropod dinosaur *Coelophysis* was collected

they were bounded by fields of wind-blown sand. Lakes formed within the red floodplains and were drained by a network of streams that coalesced to flow through a shallow, sand-filled valley in southwestern Colorado. Fishes were common and diversified in the streams that drained the lakes. Remains of the large amphibians are seldom found, possibly because of a lack of suitable habitats. Gavial-like gatorlizards were relatively common, but herbivorous gatorlizards were not as abundantly preserved as formerly. Their scarcity suggests the absence of extensive areas of low vegetation. Lake beds contain few remains of backboned animals of any kind. The rareness of fossilized bones and wood contrasts with the abundance of both in the vari-coloured clays below.

During this time in eastern New Mexico, however, alluvial fans built out into a lake have produced bones of large amphibians and gatorlizards. Such evidence as does exist indicates that the reign of the gatorlizards continued and that dinosaurs were probably not yet the dominant animals on land.

Rift Lakes

Two hundred and twenty-five million years ago, huge horizontal faults passed across thousands of kilometres, from New Brunswick in the west, overland through undulating, hilly terrain to northern Morocco and Algeria. They resembled modern faults in the coast ranges of California, and, as they moved, lands to the northwest were displaced further to the west, while to the south the crust thinned and began to rupture between what has become eastern North Amer-

Camels,
crossing dry floor
of rift lake,
northern Kenya

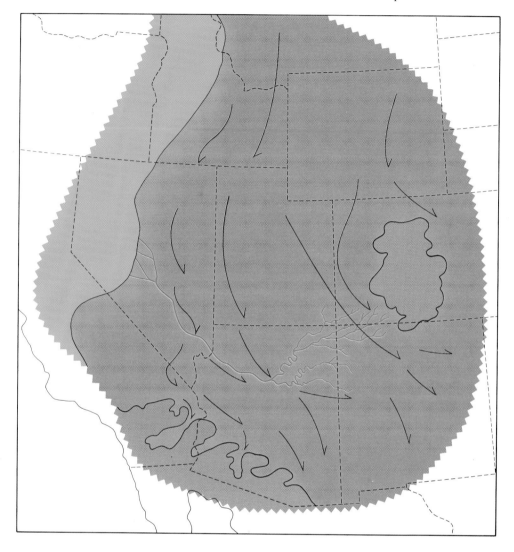

205 million years ago

A river system drains us south-west. Winds blow sands south-eastward from an enormous midcontinental sand sea (see arrows).

ica and west Africa. It was a time of earthquakes. As the edges of the forming continents foundered toward sea level, a narrow arm from the ancient Mediterranean advanced westward over northern Africa to spread into the axis of the Atlantic rift zone. By the close of Triassic time vast deposits of salt were accumulating on the floor of a shallow, 300-kilometre-wide embayment that had penetrated the rift axis for a distance of 1000 kilometres to the southwest. The salts are now preserved in the continental shelf off Newfoundland and Nova Scotia.

On both sides of the rift axis the crust broke into chains of slowly sinking blocks, which filled with thousands of metres of sediments eroded off adjacent highlands. These old rift valleys extend from Alabama to Nova Scotia. Most are now buried in the coastal plain or beneath the continental shelf. Those that formed near the western margin of

the rift zone still dominate topography in many regions along the Atlantic seaboard between North Carolina and the Bay of Fundy. The oldest known remnant of a rift valley contains bones of a swinelizard of a kind that lived in South America about 240 million years ago. Close resemblances between fish fossils – including those of lobefins (coelacanths) related to the living fossil ('old four-legs') netted in deep waters off Madagascar – and gatorlizard skeletons suggest that most of the rift valleys began to form at about the same time as red siltstones were being deposited in the great basin in the southwestern United States. The time was about 220 million years ago.

The Atlantic rift valleys were located near the northern edge of the tropical zone, where unequal warming of the hemispheres by the sun deflects equatorial rains toward the lands of the summer season. These monsoonal systems react very sensitively to changes in heating. In North Carolina and Virginia, and Pennsylvania and New Jersey, the floor of the rift valley was covered by lake and alluvial fan deposits (respectively the Cow Branch Formation, over 1500 metres in thickness, and the Lockatong Formation, 1150 metres thick). These sediments often contain varves – pairs of sedimentary layers – composed of coarse grains, deposited during the rainy season when runoff was vigorous, and fine grains, deposited during the dry season when runoff was slow. Thus each pair represents one year of sedimentation. Counting varves makes it apparent that lakes were replaced by alluvial fans every 21,000 years.

This cycle is identical in length to a cycle created by the interaction of a wobble in the Earth's axis of rotation with the slightly elliptical orbit of the planet around the Sun. As a result of this interaction, alternate hemispheres are presented close to the Sun during summers every 21,000 years. This shift produces changes in the monsoons, as a result of which ancient lakes could form in the valleys during only one phase of the 21,000-year climatic and orbital cycle. Other slight irregularities in the Earth's orbit vary cyclically, with periodicities varying between 41,000 and 413,000 years. These periodicities also affected the ancient monsoons, which in turn affected the sedimentary record. The relative durations of these cycles were similar then to those today, implying that the basic orbital dynamics of the Earth have not changed in over 200 million years. And the cycles printed in the sediments provide a precise means for measuring the passage of ancient time.

When rains were increasing and lakes were still shallow, oxygen could reach lake bottoms. Numerous bottom-dwelling organisms burrowed through the mud, searching for small bits of fallen organic matter. The varves were destroyed by their activity. Later, lakes became too deep for winds to bring bottom waters to the surface, where their oxygen content could be restored. Burrowing organisms could no longer live on the lake bottoms, and the varves were left undisturbed. When lakes were small the surrounding lands were relatively dry, and remains of only one kind of conifer have been identified. As the rains increased and the lake grew, the shores and surrounding rift mountains supported an increasingly rich flora. According to the varves, the lakes attained maximum diameters of up to 300 kilometres for only a thousand years. But they could then cover as much as 70,000 square kilometres and attain depths of more than 100 metres. The largest lakes may have been twice as large as Lake Tanganyika and Lake Baikal, which are also rift valley lakes.

When the lakes were large, their shores were lined with cycad-like plants, ferns, and giant scouring rushes, and the surrounding highlands were blanketed by seed ferns and conifers. The forests were inhabited by swarms of beetles and flies and by insect-eating gliding lizards (*Icarosaurus*). The latter resembled modern 'flying' lizards, but their ribs supported much broader and more fully formed wings. Craneflies floated over the water. Near the surface of the lake, dense populations of small aquatic organisms flourished. These provided nourishment for a series of organisms, including water beetles, fish, and small, lizard-like swimming reptiles (*Tanytrachelos*). At the top of the food chain were large amphibians (*Metoposaurus*) and gavial-like gatorlizards (*Rutiodon*).

The rains would then begin to fail. Lakes shallowed, and former outlets into surrounding river systems dried. Lakes became salty, and then dried out completely, as alluvial fans coalesced over their beds. The forests surrounding the lakes also vanished, to be replaced by scrubby bush dominated by one kind of conifer. Footprints of lizards, migrating gavial-like gatorlizards, and a variety of small, lizard-hipped dinosaurs were left behind on mud-flats. Some of the footprints (*Atreipus*) were made by primitive, bird-hipped dinosaurs. These gracile creatures often walked with their tiny forepaws alternately contacting the ground. The cycles of lake-to-alluvium sedimentation could continue for several millions of years, to be interrupted at last by changes in the structure of a basin through movements on the fracture zones along its margin.

In North Carolina and Virginia, strata (respectively the Pekin Formation, 1000 metres thick, and the Doswell Formation, 1500 metres thick) deposited on the floor of well-watered valleys contain coals, abundant plant fossils, and, rarely, reptile bones. Cycad-like plants were abundant, and one form, a true cycad, resembled a sparsely fronded tree fern. Ancient conifer cones looked much like those of modern pines, although apparently more elongate and less compact. The ground was covered with ferns. Gavial-like gatorlizards (*Rutiodon*) were common, and the remains of a very peculiar gatorlizard (*Doswellia*) are locally abundant. The latter's body form might be visualized as that of an imaginary alligator-pangolin. The armoured tail could evidently be curled forward to protect the stomach. Armoured herbivorous gatorlizards (aetosaurs) were present in the region, as well as cowturtles. These were the prey of large, wolf-crocodiles (rauisuchids). Among the smaller creatures were tiny possumlizards and small herbivorous bird-hipped dinosaurs.

Plant fossils very similar to those preserved in North Carolina and Virginia have been collected in abundance in

Camels,
crossing dry floor
of rift lake,
northern Kenya

Sonora, Mexico (Santa Clara Formation, 400 metres thick). Here impressions of leaves are also associated with coals and evidence of humid environmental conditions. The flora did not closely resemble that of the Petrified Forest in Arizona. At the time, the Sonoran site was situated in a region 800 kilometres further to the east, nearer the network of rifts from which the Gulf of Mexico would form. So far no bones have been found.

In Nova Scotia, the Bay of Fundy is famous for its huge tides, which produce a daily rise and fall of sea level of as much as 12 metres. The tides have cut the edge of the bay into eroding cliffs of reddish sandstones (Wolfville Formation, 400 metres thick), which are the same age as the lake sediments further to the south. The ancient valley floor here was occupied by dry, caliche soils and crescent-shaped sand dunes. The tips of the dunes curved away from winds that blew usually from the northeast. Large amphibians rarely occupied scattered moist areas on the valley floor. Coarse alluvial fans were spread into the valley by intermittent streams, and these contain broken, water-worn pieces of bone. The animals died in more upland regions, and the carcasses were torn apart by scavengers. The picked bones were left behind to be washed by storm-generated torrents into the alluvial fans below. A large variety of animals have been identified from the scattered fragments of bone.

As winds blew over the highlands, they were forced to rise into cooler levels of the atmosphere. Clouds formed, and the resulting rains kept the highlands well watered. Plant growth was much more luxuriant than on the valley floor, with the result that most of the animal populations were concentrated in the highlands. Here were representatives of animal groups typical of the time, such as a swine-lizard as large as a grizzly bear, cowturtles, owliguanas, and several kinds of armoured herbivorous gatorlizards. They were not spared the ubiquitous scourge of wolf-crocodiles. There were also many kinds of small plant-eating reptiles that bore spines on the back of their head and resembled very large horned toads (procolophonids). These were derived from archaic reptilian stock, which earlier in their history had subsisted on insects. The shape of the teeth in juveniles suggests that immature animals were still insectivorous.

Dinosaurs were present in some variety, including small representatives of both the lizard-hipped and bird-hipped forms. Among the former group were prosauropods, belonging to the same group of dinosaurs that occur so abundantly in strata of about the same age in Germany (*Plateosaurus*). Their forelimbs were more robust, and the

Massospondylus:
detail of animal crossing a sand dune,
Nova Scotia, 213 million years ago,
during early Jurassic time

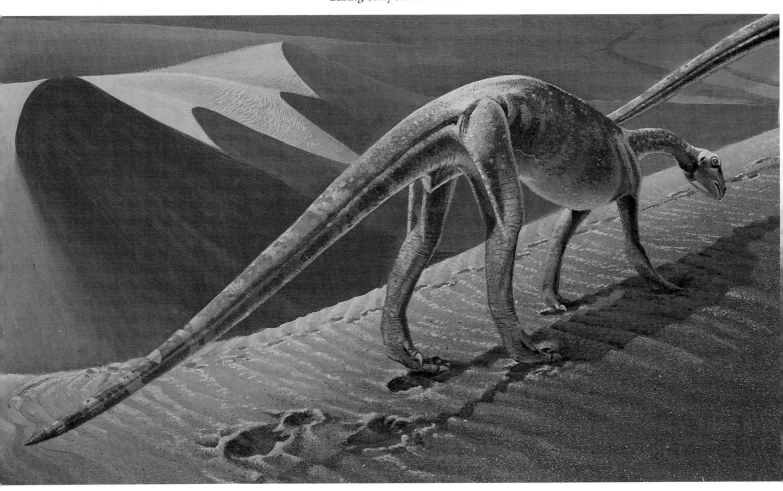

claws on the inside of the hand were powerful and recurved. The jaws were not powerfully built, but the teeth, although relatively unspecialized, indicate that the animals had become herbivores. A plant-eating diet is supported by the discovery of grinding stomach stones, analogous to the gizzard stones of some birds, preserved within the body cavities of skeletons belonging to the same kind of dinosaur found in Africa. A bipedal posture and an elongated neck enabled the animals to browse on foliage beyond the reach of other herbivores.

The abraded bones in the sands and gravels eroded from ancient highlands in Nova Scotia were derived fom animals that lived in hills and mountains. These environments are poorly represented in the fossil record, because they are usually destroyed by erosion. Conversely, deltas and low-lying floodplains are often buried and preserved for very long periods of time. Bones of gavial-like gatorlizards have not yet been found in the Nova Scotian redbeds. Perhaps their piscine prey did not occur in sufficient abundance in mountain streams.

Dry Rift Valleys

The regional climate became more arid, and lakes dwindled and vanished in the rift valleys to the south. In New Jersey, and in Massachusetts and Connecticut, the sediments (respectively the Passaic Formation, 3000 metres thick, and the New Haven Arkose, 2250 metres thick) changed to become redder and more coarsely grained. Skeletal remains are rare, but those that are preserved usually do not belong to animals that preferred stream and river bank environments. The reptiles resembling large horned toads were evidently abundant. One lizard-like reptile was related to, and closely resembled, the living New Zealand tuatara (*Sphenodon*), which has long been recognized as a living fossil. A few skeletal fragments of armoured herbivorous gatorlizards (*Stegomus*) have been collected, as well as bone scraps of gavial-like gatorlizards. There are accounts of the discovery of a hind limb of a prosauropod from northern New Jersey. Trackways are abundantly preserved, and most of the major types of reptiles are represented. The tracks of dinosaurs also indicate that some were not as small as formerly.

In Nova Scotia, sedimentation continued in an arid climate (Blomidon Formation, 300 metres thick). Fleeting alkali lakes in the valley centre were separated from the bordering hills by fields of sand dunes. Following torrential rains in the highlands, sheetfloods brought layers of mud into the playas from the northeast. Under conditions of severe drought, mudcracks formed over 40 centimetres deep. All plant materials were oxidized. The valley came to resemble Death Valley in southern California. Only a fragmentary skeleton of a gavial-like gatorlizard (*Rutiodon*) and small dinosaur tracks have been discovered.

The Reign of the Gatorlizards

The record of the last 12 million years of the Triassic in North America is preserved because the great supercontinent had begun to fragment. The uplift and subsequent breakage of the Earth's crust over the Atlantic–Gulf of Mexico rift zone ensured the preservation of enormous thicknesses of sediment in rift valleys along what was to become the Atlantic seaboard. Great rivers flowed from highlands in wetter regions of the Atlantic rift to build large deltas out into the shallow arctic sea. Crustal movements related to the eastern rifts created highlands and volcanic mountains in the southwestern United States, and sediments eroded from them to fill the great basin which they bordered. Because the major land masses of the Earth were linked in a world-continent, the same kinds of animals were broadly dispersed around the globe. Skeletal parts of a single variety of large amphibian (*Metoposaurus*) are found in the United States and India, those of a swinelizard (*Scalenodontoides*) in Canada and South Africa, and those of a prosauropod (*Plateosaurus*) in Germany and Argentina. Geography seemed to be less important in determining local associations of animals than ecology.

Following the great extinction that ushered in the Triassic Period, an equally great diversification of reptiles occurred. By the end of the 35 million years of Triassic time many new basic kinds of animals had appeared, including the dolphin-like ichthyosaurs, sail-winged flying reptiles (pterosaurs), true crocodiles, primitive turtles, tiny archaic insectivorous mammals, and dinosaurs. The character of terrestrial faunas changed, as dominance shifted decisively from paramammals to parareptiles. In sediments of late Triassic age, gatorlizard skeletal parts overwhelmingly outnumber those of all other large reptiles.

Bones of the oldest-known dinosaurs occur in strata deposited during late middle or early late Triassic time. They were initially small, but toward the end of Triassic time some dinosaurs (large prosauropods called melanorosaurs) approached 8 metres in length. These heavy plant-eaters had long necks. They avoided competition with other large herbivores in the same way that giraffes avoid competition from warthogs, by browsing on higher levels of vegetation. Dinosaurs also increased in abundance through late Triassic time. Collections of skeletal fragments from some localities in Germany and Argentina are dominated by those of prosauropods. These dinosaurs are known from North America, but their remains have so far not been found in abundance here.

Unfortunately, strata of latest Triassic age in North America have yielded very few skeletal remains of terrestrial animals of any kind. The gatorlizard host is abundant and diversified in the slightly older varicoloured clays of Arizona, at a time when the creature evidently flourished in Nova Scotian highlands as well. Scrappy material collected from redbeds of latest Triassic age is sufficient to show that gatorlizards continued to be an important component in the terrestrial fauna. This point would probably be of minor importance had the biosphere of our planet not been shaken by another profound crisis that brought the Triassic abruptly to a close.

However, while Triassic time endured, an observer familiar with life on land today would have been impressed with the strange appearance of the large paramammalian herbivores and the curious owliguanas. If he were unaware of their future importance, it is doubtful that he would have considered the primitive dinosaurs as being fundamentally different from the great variety of gatorlizards. They would have seemed like a variety rather more than usually well adapted to bipedal locomotion. This being the case, the observer would return from his voyage in time utterly amazed with the variations he had seen on a crocodilian body theme, variations that he would not formerly have considered possible. For him, and for us, the Triassic could succinctly be described as an age of crocodiloid reptiles.

Diplodocus

LATE JURASSIC, ABOUT 150 MILLION YEARS AGO

Juvenile *Diplodocus* specimens die of thirst,
leaving traces of their death struggles
in the sand, on a drought-stricken floodplain
in northern Wyoming. Small pterosaurs feed on
insects attracted to the decomposing carcasses.

The Jurassic

A Catastrophic Advent?

The easiest way to tell geological time is to check the fossil record of seashells. They are abundantly preserved and as widely distributed as the ancient oceans. A kaleidoscope of evolutionary pressures moulded their shapes into showers of differing descendants. All the major divisions of geological time were defined originally by their characteristic varieties of shellfish. Sediments deposited on land are less often preserved because they are more frequently eroded away. Skeletons are less abundant and are often destroyed or scattered by running water before burial. Thus the evolution of life on land, as a rule, is recorded in much less detail than that of life in shallow seas. When an important event occurred that affected the entire biosphere, it is usually discovered and subsequently more completely documented in shallow marine environments. Just such an event took place at the end of the Triassic Period, about 213 million years ago,

During Triassic time a group of marine molluscs (ammonites) flourished. They were related to octopuses but possessed spiral shells and inhabited open waters instead of spending most of their lives on the sea floor. They very nearly vanished at the end of the period. Other kinds of marine organisms, including lamp-shells (brachiopods) and clams (pelecypods), underwent similarly abrupt declines. Among the latter group, more than 90 per cent of the species in existence before the end of the Triassic disappeared. Those animals that were apparently more demanding in their environmental requirements less often survived. Some marine micro-organisms and possibly marine snails showed no evidence of having been greatly disturbed, and the extinctions were, in general, not so profound as the great marine extinctions that had marked the beginning of the Triassic Period.

The timing and the causes of the new extinction are controversial. One school of thought postulates that the extinctions were concentrated into two main episodes separated by millions of years. The period of faunal turnover would have been initiated by climatic changes on land and was brought to completion during the subsequent biotic crisis in the oceans. An opposing school of thought presumes a single extinction event, pointing out that an asteroid struck northern Quebec at (at least approximately) the same time that significant extinctions took place in both terrestrial and marine environments. After the removal of about 1 kilometre of rock through erosion since the time of the impact, the Manicouagan Crater still measures a very impressive 70 kilometres in diameter. Its form is now outlined by an enormous ring lake. The impact explosion is implied to have caused short-term changes in the ability of the atmosphere to retain the Sun's heat, precipitating a planet-wide biotic crisis. Great outpourings of lava occurred very soon after the event in adjacent regions of North America. The severity of the marine extinctions has been surpassed only four times during the course of the past half-billion years.

The record of life on land through the Triassic-Jurassic transition suggests an extinction on a scale similar to those in which paramammalian ascendancy was destroyed at the beginning of the Triassic. Major groups that had figured importantly in land faunas for tens of millions of years, including swinelizards (diademodonts), cowturtles (dicynodonts), owliguanas (rhynchosaurs), and gatorlizards (thecodonts), vanished from the fossil record. Other groups, such as the archaic 'horned toads' (procolophonians), lizard-like reptiles (trilophosaurs, tanystropheids), and gliding lizards (kuehneosaurs), vanished with them. The trackways made by these creatures also vanished from the geological record, and it is easy to separate late Triassic and early Jurassic strata by the reptilian footprints they contain.

Although about half of the major groups (families) of terrestrial reptiles became extinct, the remaining half survived. Within the poorly diversified surviving faunas were archaic turtles (proganochelyids), primitive flying reptiles (pterosaurs), the ancestors of modern crocodiles (protosuchians), and two groups of relatively small paramammals (trithelodonts, tritylodonts). In view of their ultimate disappearance at the end of the dinosaurian era in another great mass extinction, it is amazing that four major groups (families) of dinosaurs survived the terminal Triassic extinctions: the progressive small carnivores, anchisaurs (prosauropods), and fabrosaurs and heterodontosaurs (small bird-hipped dinosaurs). The latter two groups became particularly common immediately after the extinctions. However, primitive carnivorous dinosaurs (staurikosaurs and herrerosaurs) and possibly also the large quadrupedal prosauropods (melanorosaurs) are not known to have survived the terminal Triassic crisis.

It is difficult to evaluate the span of time during which the extinctions took place. In many areas, remains of any land-dwelling reptiles of latest Triassic age are very seldom found. The latest skeleton recovered of a species may reflect the hazards of preservation rather than an extermination. Although the die-off of reptiles of Triassic age could have

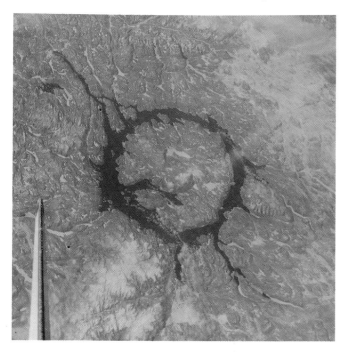

Manicouagan Crater, northern Quebec, taken from space shuttle

spanned several million years, there is evidence to suggest that it was quite sudden. The best-known exposures of late Triassic and early Jurassic continental sediments occur in South Africa. Here, the change from fossils typical of one period to those of the next is quite abrupt, occurring within a vertical distance of about a metre. If there has been no more than usual loss of record due to temporary cessation of sedimentation or erosion, then the abruptness of the change would suggest that the interval of extinctions was brief.

The ecological aspects of the extinction on land are not yet well understood. It is clear that many reptiles (swinelizards, cowturtles, owliguanas) with powerful jaws and adapted to feeding on low vegetation became extinct, while the prosauropods, with weak jaws but long necks which enabled them to feed on higher vegetation, survived. In South Africa the prosauropods were larger before. the extinctions than following them. During late Triassic time, land faunas in different regions of the globe tended to be dominated by different reptiles: in Argentina by large prosauropods, in Germany by smaller prosauropods, in the western United States by gatorlizards, and in Scotland by owliguanas. During earliest Jurassic time the faunas were both less diverse and more homogeneous. Thus animals living in eastern North America and southern Africa were not only closely related to each other, but they also occurred in similar relative abundances. This homogeneity is also apparent in the plant record, as forests of heat-resistant conifers (cheirolepidaceans) replaced older, more diverse floras. The now-extinct trees remained a very important element in Mesozoic forests for many millions of years to come.

Crocodiles were originally long-limbed terrestrial carnivores, and they became aquatic only after the extinction of aquatic gatorlizards (phytosaurs). This illustrates another interesting but idiosyncratic effect of the extinctions. The removal of gatorlizards and other large reptiles opened many ecologic niches for dinosaurian exploitation. The success of the dinosaurs then may not have been caused by their competitive superiority over their former contemporaries. Their reign was probably not inevitable and can be regarded as an age of fundamentally bipedal land animals that occurred as an interlude between two ages of quadrupedal land animals, that of the ancient parammals and gatorlizards, and that of the modern mammals (and crocodilians). The fact that the ancient and modern quadrupeds were related has caused a disproportionate amount of study to be focused on them. This would not have been the case had an age of dinosaurian ascendancy been followed by a modern world dominated more or less exclusively by birds. As will be seen, a study of the classification of extinct animals tends to underscore a random element in evolution, while a study of adaptations tends to identify recurring adaptive themes.

A Liassic Sea

The extinctions that ushered in the Jurassic Period occurred about 213 million years ago. Shortly afterward, a shallow sea spread over northern Europe under which an alternating series of dark shales and thin limestones were deposited. Beautifully complete skeletons of the dolphin-like ichthyosaurs have been recovered in slate quarries in these sediments near Lyme Regis, along the southern English coast, and near Holzmaden, in southern Germany. The strata define a subdivision of Jurassic time that has been named the Liassic after the Gaelic *leai*, flat stones. The Liassic includes the first 25 million years of the 70-million-year span of the Jurassic Period.

To the south, the Liassic sea lapped on the shores of several European islands before entering the area of the modern Mediterranean. The shallow arm, which during late Triassic time had spread into the rift zone between Africa and North America, penetrated even further to the southwest. Salt was precipitated out of its brines as far south as Virginia. Dry land linked the two continents beyond the tip of the salt sea, and mountains bordering the rift zone attained heights of nearly 3 kilometres above sea level along the Saharan Shield, and of $4^1/_2$ kilometres in the Appalachian region. Far to the west, marine waters from the Pacific crossed Mexico nearly to the modern Gulf coast.

Making Tracks in Connecticut

In the Maritime provinces and New England, lava flowed from fissures in the earth and spread into the rift valleys early in Liassic time. These solidified flows now form prominent escarpments, such as North Mountain in Nova Scotia, Mount Holyoke in Massachusetts, and the Watchung Mountains in New Jersey, all of which are of earliest Liassic age. Sheets of lava blocked streams in the valleys, and the disruption of drainage, in conjunction with increased rainfall, caused rift lakes to form once again in New England. Several major lava flows, averaging about 100 metres in thickness, repeated the damming process over the next 10 million years. Some of the perennial alkaline lakes exceeded 5000 square kilometres in area and were comparable in size to Great Salt Lake in Utah. In the Connecticut Valley,

they were sufficiently deep to prevent the circulation of oxygen to bottom waters, and grey muds accumulated (East Berlin Formation, 170 metres thick). The bodies of fish were often abundantly preserved on the stagnant lake floors, scattered among fish droppings measuring up to 15 centimetres long. In one case, the gut contents were still preserved within the body cavity of a lobefin (coelacanth). Larval lacewings and the tiny shells of clam shrimp were also abundantly preserved.

In Dinosaur State Park, Rocky Hill, Connecticut, fossil ripple marks, raindrops, and mudcracks are exposed on the surfaces of strata that were deposited near the end of a period of accumulation of lake beds (within the East Berlin Formation). More than 2000 tracks of dinosaurs have been counted at this locality. Most of the tracks belong to a carnivorous dinosaur that measured probably about 2 metres in height and 6 metres in length. It has been suggested that the trackmaker was *Dilophosaurus*, a magnificent skeleton of which has been collected in contemporaneously deposited strata in Arizona. Most of the 40 trackways of the large carnivore are directed more or less randomly across the rock surfaces. One is particularly interesting because it shows that the animal was floating in shallow water as it pushed itself lazily along on the tips of its toes. A few other tracks of smaller carnivorous dinosaurs and crocodiles have been identified.

The Connecticut Valley holds a special place in the history of dinosaurian studies in North America. The first footprint later to be recognized as belonging to a dinosaur was discovered in 1800 near South Hadley, Massachusetts. The earliest-discovered dinosaur specimen still preserved in museum collections on this continent (Peabody Museum, Yale University, specimen number 2125) was taken from rock blasted out of a well in East Windsor, Connecticut, in 1818. The first accounts of dinosaurian fossils from North America, published in 1836 and among the oldest descriptions in the world, were based on Connecticut Valley trackways. The tracks were not originally recognized as belonging to dinosaurs. It was known at the time of their discovery that giant ground birds, called moas by the Maoris, had recently been exterminated in New Zealand. Because the moas had formerly existed and were now extinct, and because many of the Connecticut tracks had an avian shape, they were at first considered as having been made by giant flightless birds. Some were visualized as being five times the bulk of ostriches, standing over 4 metres high as they waded along the strand of an ancient sea. Many different sizes and varieties of tracks were described. Groups of parallel trackways were cited as evidence of flocking.

None of the Connecticut Valley tracks was actually made by birds. Most of the footprint-makers were carnivorous dinosaurs of various sizes and kinds. Some prints belonged to crocodiles, and a few were made by small herbivorous dinosaurs. Although as a group they differ sharply from those that occur in slightly older Triassic sediments in the same region, it is remarkable that the feet of the Connecticut Valley trackmakers remained essentially unchanged through the entire 25 million years of Liassic time. As was recognized long ago, trackways are a fascinating source of information about the behaviour of ancient living animals. At Mount Tom, near Holyoke, Massachusetts, 26 trackways of one variety of carnivorous dinosaur lie parallel to each other. Because those of another variety of carnivorous dinosaur cross the parallel trackways at an angle, there was no topographic reason for the latter animals to walk in the same direction. The inference is that the dinosaurs were moving as a group, just as the hypothetical giant birds were once supposed to have done.

Although no coals were formed in the Connecticut Valley, ample evidence of plant growth on surrounding alluvial flats is preserved in lake sediments. Conifer shoots, and fronds from cycads and large ferns, have been identified. The leaf cuticles indicate that the air was humid at least locally when the plants were growing. Conifer pollen has been extracted from dark muds deposited under the lakes, and its abundance implies that conifers covered valley slopes and highlands over a broad geographic area. The climate was evidently warm and wet but was interrupted seasonally by a dry period of less than three months' duration.

After the lava flows ceased, a thick series of red sediments accumulated in the Connecticut Valley (Portland Formation, 1200 metres thick). The strata show the effects of peak seasonal rainfall. Flash floods swept down from highlands onto the 50-kilometre-wide valley floor. Their sediment-laden waters rushed through alluvial fans to spread a veneer of red sediment across the bottomlands during the waning stages of floods. They flowed around the stems of scouring rushes and tree trunks up to 30 centimetres in diameter. The trunks later decayed, and the cavities were refilled with sediment, producing natural casts. Brief periods of flooding were followed by desiccation, and suncracks formed in mudsheets draped over the stream courses. The reptilian trackmakers left behind abundant evidence of their passage on the drying mudsheets. Much of the time the rivers were dry, but water continued to flow within their sand-choked beds. In spite of the semi-arid climate, crayfish were able to survive in burrows excavated down into the wet sand.

The sandstone sheets laid down on the broad river beds have been extensively quarried for building stone. The richest Liassic dinosaur locality in North America is the Buckland (Wolcott) Quarry near Manchester, Connecticut. It has produced skeletons of three prosauropod dinosaurs. One is of a 2$^1/_2$-metre-long animal named *Anchisaurus*, which is characterized by its slender form and narrow feet. Two others belong to *Massospondylus*, a more massive animal of about the same length but with broader feet. Another skeletal fragment of a small, 3-metre-long carnivore, *Podokesaurus*, was collected from strata of the same age near Middletown, Connecticut. Further north, in Massachusetts, another *Podokesaurus* skeleton was collected from South Hadley, an *Anchisaurus* skeleton from Springfield, and a tiny, 28-centimetre crocodile (*Stegomosuchus*) from Longmeadow. No other reptilian skeletal material is known from redbeds deposited in Liassic rift valleys in the eastern United States. Strata of Liassic age that occur in northern New Jersey have yielded dinosaurian footprints, but so far no bones have been recovered. Other footprint localities have been found as far south as northern Virginia.

The Survivors at Fundy Bay

The Manicouagan asteroid impact site lies less than 500 kilometres north of Nova Scotia's Bay of Fundy. There, along the north shore of the Minas Basin, the record of Triassic life ends beneath a series of lava flows. As irregularities in the lava surface were filled with silts and sand dunes (McCoy Brook Formation, 200+ metres thick), the great sedimentary cycles produced by irregularities in the Earth's orbit indicate that the Jurassic was only 200,000 to 300,000 years old. The footprints resemble those in the Connecticut Valley, and skeletal parts collected along the base of the sea cliff indicate a peculiar mix of surviving animals.

A muddy limestone deposited in a lava-dammed lake contains the remains of fishes and small, primitive bird-hipped dinosaurs (fabrosaurs). A petrified mudflow filled with volcanic debris has produced remains of small, long-legged crocodiles and paramammals (trithelodonts) in profusion. Skeletal parts preserved in brown, stream-deposited sands are dominated by those of lizard-like animals related to the modern New Zealand tuatara (sphenodontids) and to long-legged crocodiles. The sands have yielded a prosauropod skeleton as well. Another prosauropod skeleton, belonging to *Massospondylus*, was extracted from a sand dune. The Fundy site is the richest early Jurassic locality in North America.

An Odyssey in Time

Early Liassic red mudstones and sandstones,
Scots Bay Formation, overlying North Mountain
Lavas, Minas Basin, Bay of Fundy, Nova Scotia

Sands of Scots Bay Formation,
near prosauropod dinosaur site, Minas Basin

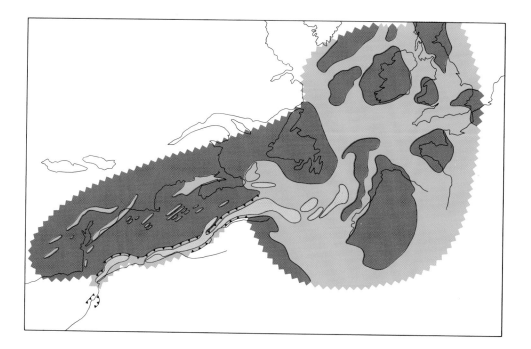

About 200 million years ago

North Atlantic rift system (see Newfoundland, British Isles, Spain). Rift basins and freshwater lakes extend across North America's eastern coastline; brines fed by marine waters from the north deposit salts along the Atlantic rift axis.

On the whole, about half of the reptilian remains from the Nova Scotian locality belonged to small terrestrial crocodiles, and at least one variety possessed caniniform ('sabre') teeth. Somewhat less abundant but probably equally varied were the lizard-like sphenodontids. Paramammals (trithelodonts) were much less common but were represented by at least two different kinds. The dinosaurs were very uncommon and very small. It was a strange fauna of 'Triassic' mini-animals, terrorized by agile little crocodiles. The fauna was, however, widely dispersed geographically. A sphenodontid, dwarf crocodile (*Stegomosuchus*), a small podokesaur (*Syntarsus*), a prosauropod (*Massospondylus*), and two trithelodonts (*Diarthrognathus, Pachygenelus*) were practically identical to forms known from southern Africa. Trackways are also strikingly similar in sediments of this age on both continents.

A Nile in a Western Desert

A series of small rivers flowed westward from the hilly roots of ancient mountains in central Colorado and New Mexico. They descended semi-arid plains of gravel in southern Utah and northern Arizona to coalesce into meandering watercourses that arced through vast marshlands in central and eastern Arizona. The river valley then turned to the northwest, diverted by a large volcanic island in the process of being assimilated into North America. It emptied into the Pacific in western Nevada. The river system was probably less than 1500 kilometres long from source to delta, but it crossed a landscape that would have been reminiscent of the Valley of the Nile. Pollen grains entombed within its red sediments bear evidence of plants of Liassic age. The regional climate was becoming drier, and giant tongues of sand were blown from the gravel fields to the east in the direction of the river valley. Marshlands were menaced by drifting sand.

The dunes were temporarily overwhelmed by seasonally flooding rivers, and sands gave way to flatbedded red silts and thin limestones laid down in sinuous oxbow lakes (Kayenta Formation, 100 metres thick). The river beds thin toward their source in the northeast but thicken (to over 350 metres) to the west, where the ancient stream turned toward the sea. The floodplains were crossed by dinosaurs, and they left behind trackways that resemble those of the Connecticut Valley – hardly surprising in view of the fact that the virtually identical South African trackways were then 8500 kilometres from New England, and Arizona, as now, was only 5000 kilometres distant.

The sediments of the fluvial wetlands contain a dinosaurian fauna that has been very poorly sampled. The first three specimens were discovered in 1942 by a member of the Navajo Nation on whose land they occurred. All three belong to *Dilophosaurus*, a carnivore allied to the late Jurassic *Ceratosaurus*, and one was a superbly preserved skeleton. The animal from which it was derived measured 5.5 metres in length, stood 1$\frac{1}{2}$ metres high at the hips, and carried its head more than 2 metres off the ground. It weighed over

Where dinosaurs trod:
three sites west and southwest of Tuba City, Arizona

Tracks of a large theropod dinosaur
(*Dilophosauripus*), upper Kayenta Formation, near
Goldtooth Spring. A wedge of dune sands from
Navajo Sandstone underlies locality to the left.

Kayenta Formation,
near *Dilophosaurus* site,
Navajo Reservation
———
Dinosaur tracks
in lower Kayenta Formation

one-third of a metric tonne. The body was more primitive than those of later carnivorous dinosaurs: the backbone was less strong, the legs were relatively shorter, and four functional fingers were retained in the hand (only three bore claws). On each side of a large and graceful head was a thin lunate crest that rose behind the nostrils to sweep to the back of the skull. The crests probably enhanced the frightening appearance of the head. The teeth are bladelike, and the reptile was unquestionably a very effective predator.

Another carnivorous dinosaur skeleton, collected in 1964, also carried paired lunate crests on its head, but it belonged to a different form of similar size. These animals did not differ greatly from slightly smaller Triassic bipedal gatorlizards and demonstrate how rapidly the role of a large predator was refilled from another evolutionary stream of reptiles after the Triassic-Jurassic extinctions. Other materials still under study indicate the presence of both smaller ('Syntarsus') and larger carnivores. A beautifully preserved skull of a small herbivorous prosauropod (Massospondylus) is about 20 centimetres long, thereby exceeding in length those of its largest southern African relatives by about 25 per cent.

There were at least two varieties of primitive bird-hipped dinosaurs. One is represented by two skeletons of metre-long reptiles. Their flanks and back were covered with a mosaic of bony scutes, or armour, from which the name Scutellosaurus is derived. The animal was remotely related to the large armoured plant-eating dinosaurs of later Mesozoic time. Remains of the other small, bird-hipped dinosaur have not been described, but they are known to have belonged to a group called heterodontosaurs. Their teeth are divided into nipping, stabbing, and chewing sections, somewhat in the manner of those of a musk-deer. Unlike in Scutellosaurus, there were excavations in the jaws for cheek pouches, and the crowns of all the cheek teeth were incorporated into a single grinding surface. The forelimbs were large, and the animals could walk on all four feet. Both of these small herbivores probably fed on relatively nutritious plant structures, such as shoots or fruits. Those preferred by the heterodontosaurs were evidently more fibrous.

Many other large reptiles were not dinosaurs. Among the more surprising survivors of Triassic swinelizard stock were abundant, large-skulled creatures the size of beavers (the tritylodont Kayentatherium). These were herbivorous. Small, archaic mammals were represented by at least three different varieties. There were two kinds of primitive true crocodiles, characterized by the fusion of bones of the upper jaw to the braincase. Protosuchus was less than a metre long and had a foxlike skull and long legs. It was probably much

better adapted to life on land than are modern crocodilians. Eopneumatosuchus was a much larger crocodile, 2 to 3 metres in length. It had a long snout suggestive of a fish diet. Isolated pieces of skull and wing bones document the presence of long-tailed flying reptiles (the pterosaur Rhamphinion) with a wing-span of $1^1/_2$ metres.

Selected sites in the wetland sediments have been quarried, and the soft blocks were dissolved in water so that remains of the smaller backboned animals could be recovered. Isolated teeth and fragments of tiny bones show that the shallow ponds and stream banks were inhabited by small amphibians, lizards, primitive turtles, the smaller surviving paramammals (tritylodonts), and at least four major varieties of archaic mammals (kuehneotheres, haramyids, morganucodonts, and triconodonts). All these animals, both large and small, and probably many others that remain to be discovered, populated a fertile river valley that flowed through an ancient desert as the dinosaurian dynasty began. They were closely related to the original survivors of a great extinction, or at least of a brusque and brutal environmental change of world-wide amplitude. Their remains are entombed in a vast Rosetta stone that will one day shed much light on the origins of the dinosaurian world.

The Empty Quarter of North America

Reflecting an increase in aridity, winds blowing from the northwest finally filled the interior regions of the river system, changing them into dry wadis (Navajo Sandstone, about 300 metres thick). Scrubby, contorted conifers followed subterranean aquifers along the wadis and from time to time were buried upright by moving dunes. Typically, these stumps were 40 to 60 centimetres in diameter and are preserved in growth position. Flooding created ephemeral ponds in the wadis, either directly or by causing the water table to rise. Thin deposits of limestone formed under these ponds, which were typically a few thousand square metres in area (although some attained diameters of several kilometres). Their tepid waters teemed with clam shrimps (cladocerans) and supported dense mats of algae. As they dried, their moist surfaces were tattooed by animals and then split by the sun. The cracks and footprints were in turn filled by the ubiquitous Liassic sands.

There are irregular worm burrows or snail trails, and tiny sprawling flecks long ago left behind by scurrying spiders, scorpions, and lizards. Some tracks have been identified as those of small, quasi-terrestrial crocodiles. However, many of the trackways belong to carnivorous dinosaurs,

Tree growing in a wadi,
Alashan Desert, northern China

———

Sand sea, Alashan Desert

like those that frequented the now-buried river floodplain. Others were of small, broad-footed dinosaurs that often touched the ground with their forepaws. A striking trackway was made by a small prosauropod as it climbed up the 25-degree slope of a sand dune.

The entire region later became more arid still, and the level of the water table in the wadis fell. Wind patterns changed, and eastern sand seas were blown into the wadis. With its headwaters cut off, the lower part of the river system in Nevada dried up and was overwhelmed with sand. Where before there were ephemeral ponds between the dunes, wind-faceted pebbles were now scattered across the dry desert floor. The same winds blew western sand fields onto the volcanic island that had become attached to the Pacific Coast. Rough terrain between the lava flows was filled with up to 225 metres of wind-blown sand (Aztec Sandstone). The dunes grew in size to heights of 33 metres and assumed the shapes of huge crescents or windrows. The area covered by dunes expanded until it filled a gigantic arc passing from southern California toward the Colorado Rockies and north into Wyoming and Idaho. Fossil dunes still extend for 400 kilometres in an east-west direction, and 1000 kilometres north-south. This is an area larger than any modern Saharan sand sea, and only part of the ancient sand sea is still preserved. It may have been twice as extensive, equalling or exceeding the gigantic sand sea of the Empty Quarter in Saudi Arabia.

The fossil dunes contain a few reptile skeletons, which resemble those found in the Connecticut Valley. A single specimen of the small fox-crocodile *Protosuchus* has been recovered. A poorly preserved skeleton of a small theropod dinosaur (*Segisaurus*) was found squatting under the steep flank of an advancing dune. Two specimens of the prosauropod *Massospondylus* have been collected, one of which was also buried in wind-blown sand in a crouching position. It died with its feet gripping the sand. Although the fossil dunes are magnificently exposed in many beautiful landscapes across the southwest, they have produced very little evidence of ancient life. Indeed it is somewhat surprising that *Massospondylus*, a rather awkwardly built bipedal reptile the size of a large man, would even venture into arid regions where plant life was sparse at best.

In southern Africa, a vast midcontinental basin extended over the whole of the subcontinent. There the river valleys dwindled to wadis that were later filled by sand seas, duplicating events in the North American west. Here *Massospondylus* is relatively abundantly preserved in red sediments deposited on river floodplains but is much less frequently found in the wind-blown sands that replaced them. The world-wide occurrence of great sand seas is a peculiar climatological feature of the late Liassic time.

The distribution of *Massospondylus* underscores a possible geographical peculiarity as well. The dinosaur also occurs in southern China, where it is known to differ from African specimens only in possessing a slightly shorter forelimb. This rather small distinction is surprising in view of the standard paleogeographic reconstructions showing a 4500-kilometre-wide ocean separating the southern continents from Asia. Such an oceanic barrier would imply that the overland route between Zimbabwe and China, by way of western Europe, was 16,000 kilometres long. According to an expanding-Earth paleogeography, the ocean is not present and the overland distance is reduced by half. The differences between land-dwelling reptiles in eastern North America and in southern Africa are comparable to the differences between Africa and Asian populations of *Massospondylus*. Perhaps the distribution of lands and ocean basins in an expanding-Earth model more accurately reflects the geography of the time.

Not all of North America was a desert. Coals accumulated beneath cycad-like trees growing vigorously on moist lowlands in the central Mexican states of Oaxaca, Vera Cruz, and Puebla. Along the coast of Michoacan, some 350 kilometres west of Mexico City, reddish sandstones were washed out between lava flows, swamps, and the sea. They are very imprecisely dated but may be approximately Jurassic. Preserved within them are trackways of at least seven varieties of dinosaurs, including four of small to large carnivores and three of tridactylate herbivores. No brontosaur prints have been identified. Large deltas along the coast of eastern Greenland, now at latitude 70 degrees north, supported mixed forests of primitive conifers and cycad-like trees. Here many varieties of ferns grew on the forest floor, and coals also formed as a result of abundant plant growth. Organic acids in the soils of both regions may have dissolved bones before they could be petrified. A few bones of the dolphin-like ichthyosaurs have been collected from marine sediments in Oregon and Alberta. A specimen of another variety of marine reptile, with a flat body, large paddles, and a long neck, was discovered in Alberta. It is the oldest record of a plesiosaur in North America.

An Opening in the Record

Near the end of Liassic time, some 105 million years ago, Honduras, Nicaragua, and the Yucatan Peninsula of Mexico were not in the position they occupy today. Instead they

Horse skull,
at base of large dune,
Alashan Desert

were located between northern Mexico and Florida, where they were linked to the southern United States through terrains now belonging to Cuba. The Gulf of Mexico did not exist, and the entire region was above sea level. Rifts that had previously fractured the old supercontinent in this region then began to spread. During an initial phase of about 10 million years, sediments were eroded from rising highlands into adjacent valleys, and continental sediments spread widely across the lands that were to border the Gulf of Mexico. Plant fossils in central Mexico and Cuba indicate that climates were at first warm and humid.

The largest fossil floras occur in Oaxaca and provide some impression of the vegetation that must have been consumed by contemporary dinosaurs. Large scouring rushes were present, as were curious forms possibly related to extinct seed ferns, which bore long, slender leaves similar to those of small banana plants. Most of the trees had foliage reminiscent of that of cycads. Unlike modern cycads, however, the trunks were long and slender, and the fronds were carried on the ends of slender, branched stems. Old fronds did not shrivel and dry as in living cycads and palms but were shed and fell to the ground entire. These trees bore male and female cones surrounded by petal-like structures that crudely resembled flowers. Their association with coal suggests that the trees must have grown rapidly and in dense stands. They closely resembled related cycad-like plants from ancient England and India, indicating that world plantscapes were generally uniform in appearance at this time. The Oaxacan flora is peculiar in that primitive conifers and ferns were not common, although both groups are abundantly represented in strata of the same age in central Honduras.

During the following 10 million years, thick deposits of salt accumulated in the widening rift, fed by brines flowing in from the Atlantic rift through a strait between Cuba and the Gulf Coast of the United States. The rift floor was then covered by one or several dead seas, the surfaces of which may have been considerably below sea level. Red sediments deposited on lands in northern Mexico were derived from lateritic soils to the north and west, indicating that climates were hot and had become more arid. Finally, 160 million years ago, normal marine waters filled the gulf to capacity as a result of a global rise in sea level. The flooding brought life into the gulf, and shallow marine sediments in western Cuba have produced bones of marine fishes and reptiles, as well as a skeleton of a long-tailed, flying reptile (pterosaur) with a wing-span of $1\frac{1}{2}$ metres.

As the Gulf of Mexico opened, another sea alternatingly spread across and retreated from a vast triangular plain that extended from Alberta and Saskatchewan, in the north, to Utah, in the southwest. At its maximum size, this shallow sea was almost as large as the modern Gulf of Mexico. It was nearly cut off from the Pacific Ocean by a long peninsula along the west coast of the continent, and few rivers drained into it from surrounding arid lands. Its isolation, in combination with hot climates, periodically caused the sea to become saltier than the oceans.

A gigantic saline lake 90,000 square kilometres in area formed in a shallow basin bordering the sea to the southeast, in New Mexico (Todilto Formation, 40 metres thick). It was maintained by sea water flowing into the basin through permeable dune sands. The lake sediments contain evidence of minor climatic fluctuations caused by the solar sunspot cycle. Small, freshwater streams flowed into the lake, and small fishes periodically migrated from them to spawn in less salty waters near the shore. The lake was bordered by a sand sea (Entrada Sandstone, 250 metres thick), which covered 210,000 square kilometres across the region where borders of Utah, Colorado, Arizona, and New Mexico intersect, north to Wyoming. It grew to the size of sand seas now in the Sahara Desert, but only to one-third the size of the sand sea in the Empty Quarter of the Arabian Peninsula. Crescentic dunes were blown by northeasterly winds into rows of sand-hills 100 metres high. The hot, arid climate was periodically interrupted by typhoons.

The shallow sea again spread over the region from the north, bringing with it waters of normal salinity. Sediments deposited on the sea floor (Sundance Formation, 50 metres thick) contained skeletons of fishes, plesiosaurs, and a broad-paddled ichthyosaur (*Ophthalmosaurus*). Remains of the same ichthyosaur are known from across arctic Canada, as well as Argentina, England, and France. Beyond the limits of the interior sea, remains of plesiosaurs have been recovered from marine sediments of this age in California, in Alberta, and on Melville Island in arctic Canada. For 30 million years, as the Gulf of Mexico opened and a shallow sea waxed and waned in the west, there was almost no record of dinosaurs in North America. The gap in the record was caused at least partly by a cessation of sedimentation in rift valleys along the Atlantic seaboard, by sedimentary conditions hostile to the preservation of bone in plant-bearing strata in northern and southern areas of the continent, and by the prevalence of harsh, arid environments across the continental interior. But dinosaurs must have been present.

Liassic land animals were a somewhat eclectic group of survivors of the terminal Triassic extinctions. Some of them would have been familiar creatures, such as lizards, turtles, archaic mammals, and probably also frogs, for the oldest

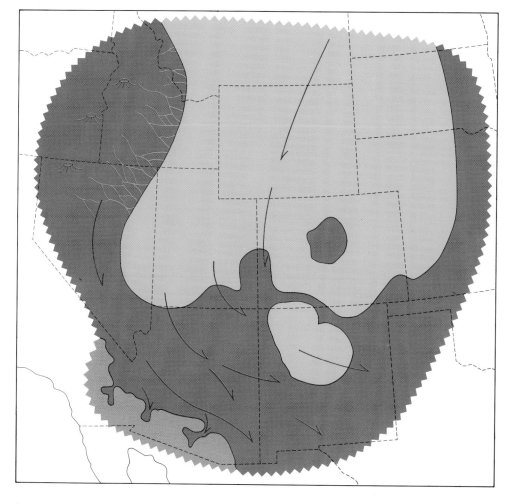

165 million years ago

A warm, shallow sea spreads south over western US states. Volcanic terrain lines its western coast; a large saline lake forms in northern New Mexico, fed by salt water seeping through sand dunes. Arrows indicate prevailing wind directions.

known true frog occurs in Liassic strata in Argentina. However, the animals that were more common and more characteristic of Liassic time were swinelizards (tritylodonts), primitive crocodiles, and prosauropod dinosaurs. Late Jurassic terrestrial faunas were very different from those of the Liassic, for they were dominated by giant brontosaurs and peculiar plated dinosaurs called stegosaurs. The record of this great faunal change is not currently documented in North America. However, because of the uniformity of terrestrial life around the world during middle Jurassic time, fossils preserved on other continents probably reflect accurately changes that occurred simultaneously in North America.

How to Build a Brontosaur

During late Triassic time a group of prosauropod dinosaurs called melanorosaurs became large, attaining lengths of 10 to 12 metres. As a consequence of their large size they also became fully quadrupedal, and unlike other prosauropods their forelimbs were nearly (70 per cent) as long as their hindlimbs. Remains of these animals have been found in England, South Africa, and Argentina. It is uncertain whether or not melanorosaurs survived the terminal Triassic extinctions. However, large quadrupedal prosauropods (or primitive sauropods) that had either descended from melanorosaurs or came to resemble them through convergent evolution were present during Liassic time in Zimbabwe. *Vulcanodon* looked like an overgrown, 10-metre-long quadrupedal prosauropod. Like its prosauropod forbears, which were tolerant of dry environments, the remains of this dinosaur were found near ancient wadis that once passed through an arid landscape. In many of its vertebral characteristics it closely resembled other prosauropods. However, the pelvis was supported by an additional sacral vertebra, and some of the claws on its feet were narrow and deep, like those of later sauropods.

The transition from prosauropod to primitive sauropod involved changes in the skeleton to support a more massive body and lengthening of the neck to crop vegetation with an economy of body movement. These changes produced a constellation of skeletal modifications that render the study of these animals so fascinating to specialists. A list of the more important of these modifications might include:

deepening of the skull,

incorporation of two vertebrae of the back into the base of the neck,

incorporation of one vertebra from the back and one from the base of the tail into the sacrum, so that five vertebral units provided stronger support for the pelvis,

lengthening of the forelimbs to support the chest in a quadrupedal pose,

broadening of the upper pelvic bone so that it could adequately anchor stronger leg muscles,

straightening of the shaft of the femur better to support the weight of the hips,

alteration of the feet which became stubby and padded as in elephants,

reduction of the claws in the hand from 3 to 1, while the claws of the foot become narrow and deep.

Prosauropod dinosaurs had a vertebral count of 10 neck, 15 back, 3 sacral, and about 55 tail vertebrae. As a result of the above changes, a primitive sauropod would have 12 neck, 12 back, 5 sacral, and about 54 tail vertebrae. The oldest true sauropod dinosaurs are represented by part of a leg in Liassic marine shales in Germany and by many bones in a large quarry of about the same age in terrestrial strata in India. The Indian remains demonstrate that by 190 million years ago sauropods had already attained weights of 6.5 metric tonnes, larger than the largest bull elephants.

Primitive sauropods lacked many of the specializations of their descendants and as a group have been termed cetiosaurs. Recently, a spectacularly rich site was discovered in the province of Sichuan in China, where most of the several different kinds of dinosaur skeletons excavated belonged to cetiosaurs. The sediments were deposited about 180 million years ago in an inland lake basin under a moist, seasonal climate. The dinosaur assemblage has been named the *Shunosaurus* fauna after the most abundant cetiosaur found at the site. Cetiosaurs of this age are also known in Argentina and Australia. Classic localities in near-shore environments in England have yielded cetiosaur bones and skeletal frag-

ments that can be precisely dated using marine shellfish. In Morocco, 170-million-year-old stream deposits have produced another assemblage dominated by cetiosaurs, and colossal footprints in the same strata record the passage of an animal that may have weighed as much as 65 metric tonnes. Within 20 million years, the largest sauropods had increased in weight by a factor of ten. Cetiosaurs dominated dinosaur faunas around the globe during this period and toward its end were showing indications of specialization and diversification.

In the course of their evolution toward a large body size, the strength of the cetiosaur backbone was increased as the height of the vertebral segments grew and the spines were enlarged to support stronger muscles to arch the back. Beginning in the front of the neck and moving rearward with time, the vertebrae were simultaneously lightened by voids (pleurocoels) that penetrated the body of the vertebra in regions of low stress. About 170 million years ago cetiosaurs gave rise to two major varieties of decendants. One lineage retained the spoon-shaped teeth, rounded muzzles, and relatively stubby tails of its forbears. However, their forelimbs were generally longer, and some grew to resemble gigantic giraffes. These may be called the camarasaurs. The other lineage was more specialized. These animals had a battery of pencil-shaped teeth at the front of a squared-off muzzle and short forelimbs, and the back was supported by ligaments suspended from tall vertebral spines in the pelvic region. The powerful tail ended in a whiplash, which may have trailed behind the animal on the ground. These were the diplodocids, among which the famous 'brontosaurus' (= *Apatosaurus*) is classified.

The reason for the division of sauropods into two evolutionary streams has not been well studied. The cetiosaurs, and the camarasaurs after them, may have preferred to feed on foliage in trees. The diplodocids perhaps became adapted to crop low, herbaceous vegetation, which they harvested by sweeping their neck over the ground like the hose of an enormous vacuum cleaner. Many similar adaptations nevertheless appeared in both groups long after their separation from cetiosaurs. In both camarasaurs and diplodocids the neck continued to lengthen through an increase in the length and number of neck vertebrae. The vertebrae in both groups were further lightened by the growth of cavities in unstressed bone. And the spines of the vertebrae at the base of the neck and front of the body in both varieties were often split into the form of a transverse 'v.' The beauty and perfection of their giant skeletons established the sauropods as enduring marvels of prehistory.

Elephant damage,
forest near Lake Manyara,
Tanzania

Little Brothers of the Giants

Representatives of another evolutionary stream of plant-eating dinosaurs had also survived the terminal Triassic extinctions. These were the bird-hipped, or ornithischian dinosaurs, descended from bipedal animals that never attained the enormous dimensions characteristic of sauropods. In spite of their relatively small size they seem not to have been very numerous, for their remains have not been found in abundance in Liassic and middle Jurassic sediments. Their presence would certainly have been less visually obvious on ancient landscapes than that of the giant sauropods, and they would not have caused large-scale damage to shrubs and trees as their powerful contemporaries undoubtedly did. Their snouts were small and pointed, indicating that they were able to pick and choose among available plant parts and were not the indiscriminate consumers of fronds, branchlets, and roots (as well as associated bugs) that the sauropods tended to be. Ornithischians can be thought of as the gleaners of the Jurassic, or as plant scavengers following in the wake of their giant companions and feeding on the more nutritious of the scraps of vegetation left behind.

The ornithischians were not protected by massive bodies as were adult sauropods. In a world inhabited by predators powerful enough to feed on sauropods and that doubtless sired numerous and vicious young, there must have been a premium attached to being fleet and well protected. It was difficult to be both; Liassic ornithischians tended to be either small and fleet or larger and armoured. As was the case with the sauropod descendants of Liassic prosauropods, some later Jurassic ornithischians differed greatly from their progenitors from a time which ended 30 million years previously. The record of intermediate forms is sparse but important. It occurs primarily in middle Jurassic strata of England and China.

Modern brushland
of primitive aspect, New Caledonia

The small, fleet ornithischians were to be a fecund source of advanced dinosaurian herbivores later in Mesozoic time. Among the most primitive were the fabrosaurs, an example of which is *Lesothosaurus* from the Liassic of South Africa. The animal was very small, measuring about a metre in length with the tail included. The forelimbs and hands were tiny, but the hind limbs were very long and give the creature the appearance of having been fleet. A highly evolved plant-eater it was not. The jaws were weak, the teeth were simple leaf-shaped structures, and there seem to have been no cheek pouches to retain plant material being chewed. *Lesothosaurus* possessed the minimal ornithischian features of a muzzel sheathed anteriorly by a horny bill, a small gap in the tooth row in the front of the jaws, and a characteristic rod-like bone that extended from the pelvis back to the anal region below the tail. Some 10 million years after the time of *Lesothosaurus*, lake beds (with the *Shunosaurus* fauna) were being deposited in central China. In them was recently discovered a nearly complete skeleton of a closely related dinosaur named *Yandusaurus*. It was larger, about 2 metres in length, and the jaws were somewhat more powerfully constructed. The eyes were enormous. *Yandusaurus* was a harbinger of more highly evolved dinosaurs to come.

The origin of the numerous and interesting groups of ornithischian dinosaurs is unclear. A few specimens of Liassic age are known, some of which had bony scutes imbedded in their skin. Their legs were short relative to the length of their bodies. Their descendants seem never to have become highly active animals, and weak jaws and dentitions imply that they never really chewed their food. Armoured dinosaurs were apparently the first ornithischians to become large, highly distinctive animals. It has been suggested that animals living in open terrain (nodosaurs) were flatbodied so that they could not be easily overturned. Armoured dinosaurs preferring bushlands or forests (stegosaurs) could rely on trunks and branches to protect their flanks, but it was necessary to defend their backs with spines and upright bony plates.

The nodosaurs were indeed squat and massive animals. They resembled the Triassic armoured gatorlizards (aetosaurs) in general body form. Bony scutes covered their head and were imbedded in their cheeks and back, while the edges of their flat bodies were protected by spines. The pelvis was expanded to form a shield covering the rump. The tail was long but did not end in a club. Nodosaurs were limited to feeding on herbaceous vegetation which, according to the weakness of their teeth, was probably not very fibrous. The oldest records date from late mid-Jurassic time in England, where they are represented by a few characteristic but isolated bones.

The shape of stegosaurs is not hard to recognize, calling to mind rather large-scale mixtures of à pangolin (scaly tailed anteater) and a porcupine. They were moderately large dinosaurs, with a small head carried close to the ground, short forelimbs, and a double row of plates or spines extending from the head along the back and terminating in a cluster of spikes at the end of the tail. *Scelidosaurus*, from the Liassic of England, is representative of the kind of animal that may have given rise to the stegosaurs. It measured about 4 metres in length, half of which was a slender tail. The scutes on its back were not yet modified into plates and spikes, but the shape of the skull was stegosaurian, and the hole for the spinal cord penetrating the vertebrae took the form of a deep oval in cross-section, as it does in true stegosaurs. *Scelidosaurus* was not a very specialized animal. It was followed in what might be representative of the next stage in stegosaurian evolution by another well-preserved skeleton from the *Shunosaurus* lake beds of central China. *Huayangosaurus* was also about 4 metres long and had a rather unspecialized skull, powerful forelimbs, and small and narrow plates along its back.

Stegosaur plates occur in 175-million-year-old strata in England. Partial skeletons of a stegosaur named *Lexovisaurus* are known from sediments about 10 million years younger in both England and France. This animal was about 5.5 metres in length, and restorations show the typical stegosaurian shortness of the front limb. It bore small, narrow plates along its back and a huge and well-supported spike over the pelvis, which was directed horizontally to the rear.

Stegosaurs did not increase greatly in size during Jurassic time, although their necks may have become slightly longer.

Another group of ornithischian dinosaurs neither appears to have been very abundant nor reached impressive dimensions during Jurassic time. These were the ornithopods, which were characterized by teeth that were enamelled only on one side. The teeth in each jaw tended to wear into a common vertical grinding surface. Pressure was maintained on the dental surfaces by muscles and ligaments in palate and cheeks. This adaptive complex seems to have been present in the small Liassic heterodontosaurs, as well as in several varieties of rather small ornithischians typical of late Jurassic time (*Dryosaurus*, *Camptosaurus*). Ornithopods, as a group, are best known from the iguanodonts and duck-billed dinosaurs (hadrosaurids), which were the dominant herbivores during the later Mesozoic.

A Note on the Opposition

Carnivorous mid-Jurassic dinosaurs have been collected at several localities in Argentina, China, England, and France, but descriptions of complete skeletons are generally unavailable. The ceratosaurs, bearing a curious notch in the upper jaw behind the nose and exhibiting a rather unspecialized skeletal anatomy, were an important group of Jurassic carnivores. Descended from Triassic forms similar to *Coelophysis*, they are represented during Liassic time by *Dilophosaurus* and *Syntarsus* and during late Jurassic time by *Ceratosaurus*. Modified descendants occur in the mid-Cretaceous of Argentina (*Carnotaurus*). Representative of the less well-known carnivores, bones of *Megalosaurus* (historically the first dinosaur to be named) are scattered through the sediments of a slate quarry in England near Oxford, where they apparently outnumber those of all other kinds of dinosaurs combined. The local environment must have constituted a death-trap for carnivorous animals.

By the end of middle Jurassic time, the actors were suitably modified from their Liassic ancestors to participate in one of the greatest paleontological dramas of all time. The setting is the mighty stage of the Morrison Formation, in the Rocky Mountains and on the High Plains of the United States.

[4]

Apatosaurus

LATE JURASSIC,
ABOUT 150 MILLION YEARS AGO

Two *Apatosaurus* individuals cross a
monsoonal rain pond in Colorado,
which is lapping around tail vertebrae
of a *Camarasaurus*. A lone *Stegosaurus*
browses at the edge of an open cycad-
conifer forest in the background.
The Moon, much closer to Earth
than now, lacks later, meteor-
induced crater Tycho.

A
Late Jurassic
Plain

About 160 million years ago volcanic islands from the Pacific were crushed against North America, creating rugged coastal ranges in central Idaho, eastern Nevada, and western Utah. The level of the oceans fell slightly, and the sea that had covered the high plains and eastern Rocky Mountains dwindled to a shallow arm extending inland from the Pacific to the Dakotas. The old sea floor drained and was slowly covered by a thin mantle of varicoloured, indistinctly stratified sediment, which spread from the western ranges toward the retreating sea to the north. Called the Morrison Formation, this sedimentary sheet still extends for 1500 kilometres from New Mexico to Canada, and 1000 kilometres from Idaho to Nebraska. Measuring well over 150 metres in thickness in the west and thinning to 60 metres in the east, it accumulated over an interval of 6 million years. Each metre of sediment thus records a span of about 60,000 years, which is 10 times longer than the written record of human history. Depositional rates were very low, and plant and animal remains were exposed to a high risk of destruction by soil organisms and the effects of the elements.

An anastomosing network of seasonal streams entered the lowlands from Idaho, carrying particles eroded from older Jurassic sand dunes and volcanic lands further to the west. A great alluvial fan formed on the southwestern edge of the lowlands in southern Utah. Trees grew along the higher part of the radiating watercourses, but the streams dwindled as they descended toward the lowlands from the forests. Another alluvial fan descended to the north through fields of drifting dunes, from ancient hilly terrain in western New Mexico. To the northeast, particles of older sediments were scoured from a low plateau in Nebraska and blown by southerly winds into a sand sea that extended at least 160 kilometres across the lowlands into the region of what is now the Black Hills of South Dakota. The sky above the arid plains must often have been filled with wind-blown dirt.

Although the climate was semi-arid, the lowlands were too vast and too close to sea level to be well drained. Very minor warpings of the earth's crust produced visually imperceptible swells and basins, creating large areas of internal drainage within the flatlands. Swamps and soda lakes were widespread across Utah, Colorado, and Wyoming. High evaporative losses caused the lakes to shrink seasonally, and mudcracks and dinosaur footprints indicate that even the deeper areas often dried out completely. Sediments washed into the lowlands were predominantly reddish in colour, but in poorly drained areas they were reduced to shades of grey by decaying plant material. Sand fields formed even around areas that remained perennially wet. One can visualize a seemingly boundless plain of alluvium, stretching to a hazy horizon through which the sun passed on quiet dawns and hot, windy sunsets. During the long dry season the furrows of meandering, interconnecting wadis were the most prominent topographical features. Dunes were blown from sands left by evaporating streams as they attempted to cross the former floor of the sea, and great dust devils wandered in pairs across the plains at noon.

After a few million years the dry seasons became shorter. Sand fields were no longer fed by blowing sand, and old dunes were 'frozen' by the tenacious roots of plants. Streams flowed more vigorously from the highlands to the southwest, carrying gravels along their beds far across the flatlands. Some of the larger rivers ran during the entire year. Lakes were fed by a greater abundance of water, but some were also drained as a result of the development of a more mature, integrated system of watercourses. To the north, the edge of the sea was filled by networks of minor

Gravel beds of intermittent stream descending from
Lapurr Range toward Lake Turkana, northern Kenya

Floodplain soils heavily ploughed by passing
feet of large mammals, Luangwa Reserve, Zambia

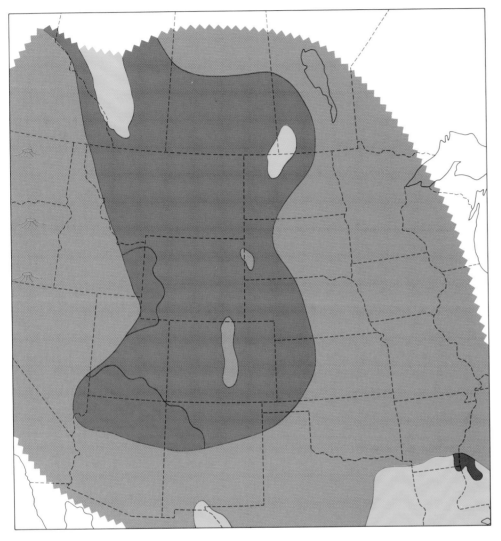

160 million years ago

Western North America's vast, semi-arid interior plain. Seasonally flooding streams deposit sand and gravel fans along its southwestern edge. To the north, the arm of a shallow sea is retreating; to the east, a large, semi-permanent lake is forming in an interior basin left behind by the sea.

deltas in which coal swamps formed. Sandy coastal beaches were displaced north into Canada. Even there the sea became so shallow that its floor was disturbed periodically by waves created by passing hurricanes. The sea had already withdrawn from the east, leaving behind shallow freshwater lakes up to 100 kilometres wide in North Dakota. Increasingly abundant remains of plants and bones indicate that life began to prosper across the broad lowlands.

The seasons nevertheless continued to exert potent climatic stresses. Some lakes still became alkaline during the dry season, when many ponds and swamps went dry. The flatlands sloped so gently to the sea that deep, permanent stream channels seldom formed. After torrential downpours, sheetfloods swept down watercourses, inundating adjacent lands with catastrophic rapidity. The absence of stratification in these areas and throughout the lowlands suggests that when the soils were wet they were ploughed

by the passing feet of thousands of giant dinosaurs, and the plant remains they contained were usually destroyed by oxidation through repeated exposure to the air.

The plant life of the great interior flatlands is thus poorly understood. Such fragmentary materials as are preserved indicate that many of the more abundant trees and shrubs belonged to groups that have since become extinct, and their growth forms are unknown. Vegetational reconstructions are largely hypothetical, although they will surely be improved with study. Enough rain probably fell on the highlands to the west and southwest to support forests composed of many kinds of conifers. Some were related to the podocarps and araucarians that form forests today in New Zealand, while others were allied to redwoods. The crowns of the trees were probably intermediate in shape between those of living conifers and hardwoods and were typically covered with dense clumps of small, bladelike leaves.

Modern conifers of primitive aspect
(araucarias), dominating highland
vegetation, New Caledonia

Giant fern tree in moist forest,
New Caledonia

Tree ferns grew in disturbed areas along stream courses where light could reach the ground, and many smaller fern varieties grew among the tree bases on hilly slopes. Tree trunks tumbled into alluvial fans after heavy rains, and the small, light cones were washed far out into the flatlands below. In the appropriate season tenuous clouds of pollen were blown from the trees far across the basin.

The appearance of the flatlands was affected greatly by the seasons. They could be vast, arid plains where river beds dried up and ponds of water became too saline to drink. Oases where the water remained sweet were lined with gnarled evergreens. Small (2.5-metre-long) crocodiles lazed on their shaded, dusty surfaces, eyeing turtles ambling slowly along the banks. A ploughed surface of grey dried mud, broken perhaps by a heap of dry dung, an abraded branch, or a giant bone, stretched away to the horizon. There were no stones, although an occasional turtle shell that had not been stepped on may also have relieved the monotonously ploughed mud-flats. Lines of scraggly conifers crossed the plains, which were either green or skeletonized, depending on whether the streams that nourished them had erratically changed their courses during a flood. Away from the oases and dry stream courses the earth supported low, pineapple-like 'cycads' (bennettitaleans), crowned by small thorny leaves that from a distance gave the appearance of thin fields of sagebrush. Within the reddish soil, tubers lay dormant.

And then the rains came. Clay pans filled with water, and waving fronds of ferns emerged in dense carpets among the sprouting cycads. Fungal brackets grew from water-soaked logs lying among the ferns, and clumps of green moss spread over the bleached surfaces of giant bones. The dry plains were transformed into lush green prairies, dotted with swamps and broad blue lakes. Populations of snails and insects spawned in the rains, providing food for frogs and some of the most ancient salamanders known. These in turn whetted the appetites of lungfish, emerging from the burrows in which they hibernated, and of voracious ancestral bowfins that swam with the rising waters from small, seasonal oases. Lizards, some of which had already acquired the ability to shed their tails when pursued, and many species of tiny, insect-eating mammals found abundant food and shelter beneath the prairie ferns. Groups of shy, metre-long fabrosaurs fed on small fruits and soft, young, leafy shoots. Small land animals in turn fell prey to chicken-sized dinosaurian carnivores and small, long-legged, running crocodiles. The most placid, and obvious, of the smaller animals were probably turtles, some of which had shells that were 60 centimetres long.

Because of the dry season, populations of small animals could not become as dense as they might have if plants could have continued to grow throughout the year. With a few local exceptions, their bones are not abundantly found in the sediments of the ancient flatlands. Perhaps many were also destroyed through trampling or weathering during the dry season. Their remains seldom occur in concentrations on stream beds as they often do in younger strata, and fewer varieties are known. The low food value of the coarse vegetation protected the plants from small, energy-demanding herbivores. A heavy thorn measuring 3 centimetres across the base and 2 centimetres in height is a measure of the need for protection from the larger browsers.

In central Montana and southern Alberta soils were moistened by ground water moving slowly toward the sea and were thereby capable of sustaining the growth of lowland forests. These forests may have been rather peculiar in appearance. They apparently contained several varieties of trees related to the maidenhair tree, or ginkgo, which has survived in the orient to modern times. These ancient ginkgos may have borne straight branches set at characteristically acute angles from the trunk, as do their modern relatives. Their fanlike leaves were divided into several lobes by deep clefts. Like most temperate hardwoods, modern ginkgos shed their leaves seasonally. There were many different varieties of cycad-like plants, but all seem to have been rather uncommon. In order to compete with other trees, they may have grown taller than their pineapple-stemmed relatives to the south. Blossom-like male cones have been preserved, and the bases of some of their coarse but fernlike fronds appear to have had an abscission zone. Like the ginkgos, the cycad-like plants may too have seasonally shed their leaves. Other peculiar 'seed fern' trees (caytoniales) possessed leaves containing a rudimentary network of fine veins in a manner foreshadowing the leaves of broad-leaved trees. The podocarps and araucarias probably retained their bladelike leaves throughout the year.

If the forests were partially bare during the dry season, they would have been protected from high water loss, as well as from heavy browsing by migrating herbivores when the fern prairies to the south had dried up. In some regions, however, conditions were moist enough for coal swamps to form, suggesting that plant growth could be luxuriant. This is somewhat paradoxical in that the wood of living ginkgos, cycads, podocarps, and araucarias does not conduct water as efficiently as that of typical modern trees. These primitive forms now grow so slowly that it is difficult to imagine them producing coal under seasonally dry climates. Modern forest trees grow more rapidly under artificial conditions when the carbon dioxide content of the air is enriched. Could higher growth rates have been sustained in ancient trees because of a relatively enriched carbon dioxide content of the Jurassic atmosphere?

During Morrison time, lakes were larger and more permanent features of the inland plains in southeastern Colorado. Lake-deposited sediments and dinosaur trackways are particularly well exposed in one site within the valley of the Purgatoire River. Here the configuration of ripple marks indicates the former existence of a body of water 100 kilometres or more in diameter. There is some evidence of shoreline fluctuations during periods of drought, but the waters were never more than slightly alkaline, for they supported permanent populations of gilled snails and fish. Prevailing winds were from the north and northeast, from the interior of the continent. Sediments along the shoreline were limy, indicating that the lake was clear and not fed by large, sediment-laden streams. However, sheets of mud and sand were sporadically laid down in the wake of storms.

Approximately 100 dinosaur trackways have been mapped along an east-west–trending shoreline dotted with clumps of scouring rushes. Two-thirds of them consist of tridactylate prints made by bidepal dinosaurs. A few bear impressions of talons, indicating the passage of large flesh-eaters, but most were made by somewhat smaller herbivores. The latter frequently moved in small groups of 3 to 5 animals. Tracks of diminutive bipeds are rare, probably because they tended to avoid exposed environments near the lakeshore. No prints belonging to quadrupedal stegosaurs have been identified.

Sauropods are represented by tracks made by broad- and narrow-footed varieties, many of which were subadult. The animals tended to associate with others of their own kind and size category, in groups of 5 or 6 animals which walked behind each other more or less in a line. The trackways skirt the lake margin, and none bears directly toward deeper waters. These sauropods were apparently not attracted to the lake and preferred not to enter it.

The excavation of skeletons of Morrison dinosaurs began when the American west was young, and it is easy to feel the pulse of the frontier in the writings of early collectors. The names they gave to the skeletons they collected, and to the animals to which they belonged so long ago, have often become modern household words. They are easily learned by preschoolers, usually to the amazement of their elders. Most were coined late in the 19th century by Othniel Charles Marsh of Yale University, when he was engaged in intense rivalry with Edward D. Cope of Philadelphia. Both men had great talents and great shortcomings, and much has been written of their struggle to be the first to collect and name exotic skeletons preserved in the fossil fields of the west.

Most of the names of Morrison dinosaurs proposed by Cope were either based on material too fragmentary to be identified or appeared in print after a name given by Marsh

A Late Jurassic Plain

Elephant family by lake, Luangwa Reserve, Zambia

Marsh Quarry: low in Morrison Formation,
north of Canyon City, Colorado.
The original specimens of *Ceratosaurus* and
Diplodocus were collected at this site.

to the same animal had already been published in a scientific journal. Because of these circumstances, they cannot be used as valid biological names. Had this not been the case, children would possibly now be learning *Amphicoelias* (?for *Diplodocus*), *Epanterias* (for *Allosaurus*), *Hypsirophus* (for *Stegosaurus*), *Symphyrophus* (?for *Camptosaurus*), and *Tichosteus* (?for *Dryosaurus*). Marsh's contribution to our knowledge of dinosaurs was ultimately greater than Cope's, and in recent years two additional Morrison dinosaurs have been named after him (*Othnielia, Marshosaurus*).

Skeletal materials representing a minimum of many hundreds of individual dinosaurs have been collected from exposures of the Morrison Formation that surround the eastern ranges of the Rockies. Here, once deeply buried Morrison strata were bent to the surface by crustal movements that formed mountains many tens of millions of years later. Sediments had accumulated very slowly on the old Morrison plain. Even the most solidly constructed limb bones of giant dinosaurs that died on it were probably completely destroyed by the elements within a few decades. Skeletons could nevertheless be preserved when they were buried rapidly, and this often occurred in stream channels that were partly or completely plugged with sediment after heavy runoff, and then abandoned.

At least two major kinds of skeletal accumulations can be identified. One, possibly the result of sheet flooding, swept together skeletons of animals that had died over a wide area during the dry season. It represents a broad sampling of the dinosaurs within a region and can be recognized by the presence of all three major varieties of sauropods. Another kind of skeletal accumulation contains only one or two varieties of sauropods, representing a more local sampling. These associations reveal something of the ecological preferences of Morrison dinosaurs.

The upper parts of dinosaur skeletons were often damaged before burial, and when they were swept along in currents they frequently broke into fragments containing bones of differing surface areas and densities. For example, the light necks of sauropods were often separated from the remainder of the skeleton, and frequently only the relatively heavy feet, limbs, and tails were left behind. The skeletons of sauropods are seldom complete, nor are those of smaller forms, although the remains of the latter are preserved with a degree of completeness unusual in dinosaur-bearing sediments of other ages. Because in their totality the fossils represent the largest sample of dinosaurs known from a single sedimentary unit on Earth, it is not surprising that skeletal segments missing in one specimen may be preserved in another. The shapes of these dinosaurs are

among the most familiar and yet most interesting of all animals. Because of the relatively large amount of information available on the fauna as a whole, the characteristics of each of the major varieties (genera) known are listed below.

It must be remembered that these creatures have long since died, and some data such as length, height, and weight that are obvious in living animals can no longer be measured directly. Lengths and heights were estimated using combinations of skeletal pieces belonging (it is hoped) to the same variety of dinosaur. In some cases where the body form is very incompletely known, these measures were extrapolated from related animals. A mathematical relationship between the dimensions of the limb bones and body weights in living mammals and birds was applied to relatively complete sets of dinosaur limbs. The calculated weight was then scaled to determine the weights of the largest and smallest dinosaur specimens of that variety known. Some of the lengths, heights, and weights are only gross approximations, made in order to convey some notion of the size of the animal. The behavioural characteristics of the dinosaurs, which would also have been obvious when they were alive, must now largely be inferred from the form of their skeletons. When these animals were alive, some 150 million years ago, their brains were probably relatively smaller than those of later dinosaurs and most modern reptiles. The animals probably possessed a limited range of rather stereotyped behavioural patterns sufficient for existence in a more primitive, simpler biological environment.

Plant-Eating Bipeds (Fabrosaurs)

Nanosaurus: length 0.5 metre, height at hips 0.18 metre, weight 0.75 kilogram. What is known of the anatomy of these diminutive dinosaurs indicates that they were relatively fleet reptiles that lacked cheeks and probably fed on soft plant tissues. Their remains seem to be associated with sites where the remains of lizards, mammals, and other small creatures also occur. This suggests that they too preferred areas that were sheltered by reeds, fallen vegetation, or other natural protection. Even tinier specimens of herbivorous dinosaurs have been collected in western Colorado which, although adult, evidently stood only 0.16 metre high at the hips.

Othnielia: length 2.4 metres, height at hips 0.7 metre, weight 19 kilograms, (for the smallest known specimen: length 0.5 metre, height at hips 0.14 metre, weight 0.16 kilogram). These animals appear to have been larger than, but closely related to, *Nanosaurus*. Like *Nanosaurus* they apparently lacked cheeks and possessed primitive, leaf-shaped teeth borne in a rather open row along the jaws, as in modern herbivorous iguanas. Their skeletal parts have been found either alone or associated with those of other small animals that presumably frequented sheltered environments.

Plant-Eating Bipeds (Ornithopods)

Dryosaurus: Length 3.2 metres, height at hips 1.0 metre, weight 100 kilograms (for the smallest known specimen: length 1.1 metres, height at hips 0.4 metre, weight 4 kilograms). The head of *Dryosaurus* was rather rabbit-like, with a squared-off snout suitable for browsing on flat surfaces. The teeth were inset toward the tongue, suggesting the presence of cheeks, and they were worn to form a single but irregular chewing plane. The neck was not unusually flexible and the body was long, so that the animal may have tended to pivot its entire body vertically about the hips to bring the head to the plant material desired. The strength of the teeth suggests that the animal chewed its food rather thoroughly to speed digestion so that it did not have to carry a large ballast of unassimilated food. The legs of *Dryosaurus* were long, and the first toe had been greatly reduced in the same way that lateral toes in the feet of modern hooved mammals are reduced. It obviously depended on speed for protection against carnivores. The animal was evidently not abundant, and its remains may show an association with those of *Diplodocus*. Perhaps it scavenged plant materials uprooted or broken and left behind by the feeding giants. However, at one site in Tanzania a dense accumulation of dryosaur skeletal fragments indicates that a large number of the animals met their death suddenly, perhaps because of a flash flood. Such a concentration would be unlikely unless the animals occasionally herded.

Camptosaurus: length 5.2 metres, height at hips 1.4 metres, weight 1000 kilograms (for the smallest known specimen: length 2.1 metres, height at hips 0.6 metre, weight 76 kilograms). *Camptosaurus* resembled *Dryosaurus* in general body form but was somewhat larger and much more robust. There are many differences in detail between the two animals. The snout was longer in *Camptosaurus* and ended in a small, expanded beak that gives the impression of being suited for selective browsing in leafy foliage. The teeth formed a single grinding surface on each jaw, as in *Dryosaurus*. Vegetation must have been retained in the mouth by cheeks and rather well chewed. The neck, although short, was

highly flexible, allowing the head to be held in many positions without moving the body. The back was broad and contained one more vertebral segment than in *Dryosaurus*. In feeding, the animal may have often rested on its forelimbs, for the wrist and hand are very solidly constructed. Its hind legs were also powerfully constructed, and the tail dropped rapidly to the ground behind the hips.

There seems to have been some ecological separation between *Camptosaurus* and the sauropods. Perhaps they frequented areas where plant growth was too sparse to sustain the larger animals. In at least one site they occur in some abundance with *Stegosaurus*, together with a few specimens of the sauropod *Camarasaurus*. Fragmentary remains of a giant camptosaur have been collected, indicating an animal over 8 metres long and weighing nearly 4 metric tonnes. The name *Camptosaurus amplus* has been proposed for it, should it prove to be different from the smaller variety, *Camptosaurus dispar*.

Plated Dinosaurs (Stegosaurs)

Stegosaurus: There were at least two species of Morrison stegosaurs. One had long legs, a high rump, and a tail ending in four pairs of spines (*Stegosaurus ungulatus*); the other had shorter legs, a lower rump, and a tail ending in two pairs of spines (*Stegosaurus stenops*). The rump was not so magnificently developed in immature stegosaurs, nor in

Cope Quarry: high in Morrison Formation, north of Canyon City. In this quarry (at base of small reddish hillock in middle distance) were found original specimen of *Camarasaurus* and several giant Morrison dinosaur specimens.

most of the stegosaurs known from other continents. The two North American species differ somewhat in size: *Stegosaurus ungulatus*: length 4.7 metres, height at top of hip plate 3.6 metres, weight 6400 kilograms; *Stegosaurus stenops*: length 4.5 metres, height at top of hip plate 2.4 metres, weight 3700 kilograms (for the smallest known specimen, of unknown species: length 1.6 metres, height at top of hip plate, if present, 0.9 metre, weight 163 kilograms).

A row of vertical plates along the arched backs of stegosaurs, and the presence of paired spikes on their tails, render them the most easily recognized of all dinosaurs. The plates were long considered to have been arranged in an alternating series along the back and to have served as defensive armour. It is, however, possible that they formed a single row and that their tips were inclined alternatingly to the left and right along the back. Although firmly imbedded in the animal's back, the plates were very lightly constructed and filled with blood channels. The surfaces were evidently not protected by horny sheaths but were covered only with skin. They were well placed to dissipate heat convectively into a wind blowing from any direction. By controlling the blood flow to the plates, stegosaurs could thus have controlled their heat loss. Undoubtedly, the plates were used as organs of defence *in extremis*.

Stegosaurs were peculiar in many other ways as well. Their snouts were long and pointed, suggesting that the animals preferred to snip off selected pieces of vegetation. Because of the length of the powerful forelimbs and the brevity of their necks, they probably preferred to feed on vegetation growing within a narrow vertical range, well above the surface of the ground. It is easy to visualize them passing their beaks between the woody stems of mature cycad fronds to pluck tender shoots sprouting from the centres of the low crowns. Their teeth were very small, as were their heads, but their bodies were enormous. The stomach was probably a huge, mobile fermentation vat where bacterial action changed coarse plant parts into food that could be digested. Perhaps the plates were necessary to rid the animals of the heat of digestive fermentation.

Stegosaurus is famous for a cavity in the vertebrae of the hip region through which the spinal cord passed, which was expanded to a volume about 20 times that of the brain cavity. Many have mused over the notion that the animal may have possessed an 'after-brain.' The brain of an octopus is a bit more decentralized than in most animals, but a degree of separation amounting to the distance between the head and hips of a stegosaur would be quite extraordinary. If the nervous tissue postulated to fill both areas were combined, the total amount would be equivalent to the size of the brain in primitive mammals, after correcting for the effect of the reptile's greater body size. This would in turn imply an anomalous degree of cleverness, far superior to that of contemporaneous plant-eating dinosaurs. Birds often store sugar in special tissues that lie next to the spinal cord in the hip region. The chamber in the hip region of stegosaurs was probably not entirely filled with nervous tissue.

Stegosaurs tended to live well away from watercourses in regions where exposure to heat and to predators was high. The plates and spikes provided some protection from carnivores, and their throats were shielded by a mail-like coat of small bony discs imbedded in the skin. Isolated bones and skeletal fragments of stegosaurs are often found in quarries dominated by sauropod skeletons, but the animals are seldom abundant there. A great accumulation of stegosaur bones in Tanzania is, as in the case of a similar occurrence of dryosaur skeletons, suggestive of herding behaviour.

Plant-Eating Quadrupeds (Sauropods)

Haplocanthosaurus: length 14 metres, height at hips 3 metres, weight 7000 kilograms. The skull and limbs of *Haplocanthosaurus* are unknown, but this relatively small sauropod can be recognized by its simply constructed backbone. The spines along its back were undivided, as in primitive cetiosaurs of early and middle Jurassic age. It was evidently a living fossil for its time. The animal was neither large (for a sauropod) nor common, but it was broadly distributed.

Apatosaurus: length 23 metres, height at hips 5 metres, weight 42,500 kilograms (for the smallest known specimen: length 4.5 metres, height at hips 1.1 metres, weight 258 kilograms). *Apatosaurus*, together with *Diplodocus* and *Barosaurus*, belonged to a group of sauropods called diplodocids. It will be recalled that that group was characterized by sheep-like heads with pencil-shaped teeth, by a tendency for the neck to become quite long, and by a backbone suspended from ligaments attached to tall vertebral spines over the hips. The tail ended in a long, slender 'whiplash.' All three forms appear to have fed on low vegetation, sweeping their necks in great arcs over the ground. The forelimbs were shorter than the hind limbs, and the necks were marvels of lightness. Studies of weight and stress distribution in the skeletons suggest that the animals could raise their forelimbs and pivot their bodies around the hind limbs with little effort. They represent a superb adaptation to sustaining a maximum volume of living tissue on a minimum of food per unit body weight, which through their long necks and

A Late Jurassic Plain

Morrison Formation, near Fruita, Colorado: reddish sediments, deposited on well-drained floodplain. An *Apatosaurus* skeleton was collected from quarry in centre.

Morrison Formation, near Morrison, west of Denver. South Platte Formation of Dakota Group forms ridge at crest of hill. *Apatosaurus* and *Stegosaurus*, collected first from sediments originally deposited on moist floodplain, are now exposed on hillside beneath Dakota sandstones.

Apatosaurus:
detail of animal
wading into monsoonal
pond, Colorado,
150 million years
ago, during late
Jurassic time

(RIGHT)

Diplodocus:
detail of carcass
lying on a dry
lake bed, Wyoming,
150 million
years ago

cranelike body movements they cropped with an economy of effort.

Apatosaurus was a very large and heavily built animal. The ponderousness of its appearance was accentuated by the fact that it carried its belly low, clearing the ground by only about a metre. No skull has ever been found attached to an *Apatosaurus* skeleton, although one was collected near a well-preserved specimen and may have belonged to it. If so, the head was rather low and broad. The bulky neck was 6 metres long in large animals. At the base of the neck and front of the body the vertebral spines were forked, possibly in order to support the neck better as it moved over the ground. The massive spines above the hips, from which the backbone was suspended, were tall and undivided. The tail ended in a very long whiplash. The elephantine forefeet bore a powerful claw on the thumb, which may have been used to push the animal to the side as it turned about its hind limbs. The three blunt inner toes of the foot also bore

claws. *Apatosaurus* seems to have been a rather solitary creature, and quarry data suggest that it tended to avoid habitats preferred by *Diplodocus*. Both animals were ground feeders and evidently sought different kinds of fodder. Of all the larger herbivores, *Apatosaurus* may have been ecologically most closely associated with *Stegosaurus*.

How this reptile received its name points to an interesting discrepancy between the popular and biological use of words. In 1877 Professor Marsh gave the name *Apatosaurus* (illusionary lizard) to an immature skeleton of a sauropod discovered near Denver, Colorado. The skeleton of a massive animal collected on Como Ridge near Medicine Bow, Wyoming, was named *Brontosaurus* (thunder lizard) by Marsh in 1879. An illustration of the probable appearance of the skeleton of *Brontosaurus*, and the first published skeletal reconstruction of any sauropod, appeared in 1883. It provided a visual impression of what the huge animal may have looked like and was widely reprinted. The first

skeleton of a sauropod ever to be mounted, also of a *Brontosaurus* specimen collected near Medicine Bow, was placed on exhibition in the American Museum of Natural History in New York in 1905.

The impact of *Brontosaurus* in the public mind was thus considerable, and other skeletons of the same animal were later mounted at the Chicago Field Museum of Natural History, at Yale's Peabody Museum, and at the University of Wyoming. Until about 1910, when larger dinosaurs were discovered in Tanzania, *Brontosaurus* was generally considered the biggest land animal that had ever lived. But by 1903 another dinosaur specialist came to the undesired conclusion that *Brontosaurus* was in reality a mature *Apatosaurus* and that the latter name should be used for the animal because it had been proposed earlier. His decision was supported by later studies. However, it was through the name *Brontosaurus* that the image of a sauropod dinosaur had captured the imagination of the public, and almost everyone who knows the name 'dinosaur' also knows what a brontosaur looked like. The word has passed into common North American English usage.

Diplodocus: length 23 metres, height at hips 4 metres, weight 16,000 kilograms (for the smallest known specimen: length 4.7 metres, height at hips 0.8 metre, weight 118 kilograms). *Diplodocus* was a large but skinny sauropod. The skull was narrow, and the slender neck was at least as long as in the largest specimens of *Apatosaurus*. Almost all the spines of

the neck and back were forked, suggesting great flexibility. The tail was held well off the ground. Heavy wings of bone on the sides of the vertebrae at its base indicate that the tail could be rapidly and powerfully pulled from side to side. Many *Diplodocus* tails show evidence of damage to the vertebrae at midlength, which is suggestive of overuse. The whiplash may have been an effective weapon of defence. Only the inner digit of the forefoot and the three inner toes of the hind foot bore claws. *Diplodocus* seems to have been a somewhat more social animal than *Apatosaurus*, for its remains are found more often in association with those of other sauropods and smaller, bipedal plant-eating dinosaurs.

A great accumulation of *Diplodocus* bones was excavated in 1934 at a famous locality known as Howe Quarry, located near Greybull in northern Wyoming. According to a detailed quarry map in the American Museum of Natural History in New York, pieces of at least 19 subadult animals were found. They ranged in length from 9 to 18 metres and in weight from 1 to 8.5 metric tonnes. The young sauropods had evidently died, perhaps because of a drought, and their carcasses had partly rotted. Their bodies were then swept up in a flood, partly broken apart, and piled, pell-mell, together in a muddy backwater. The weakest ligaments were broken, so that limbs were torn from bodies and vertebrae of the neck were often separated and scattered. This circumstance suggests that there was no heavy tendon passing from the back between the forks of the cleft vertebral spines along the neck and that each neck vertebra was

Bone webbing in neck vertebrae of a rather small *Camarasaurus*, exposed in cross-section on blocks of Morrison sandstone, near Rabbit Valley, western Colorado

connected largely by muscles to adjoining vertebrae. Many tail whiplashes were preserved intact, as well as pelvises and limbs. Isolated patches of skin casts were found scattered among the bones. The skin was covered with small bumps the size of the head of a large pin, and there was no evidence of different patterns of bumps in skin from different parts of the body. A small fragment of a cast of sauropod skin has been reported in only one other locality, in southern England. A pile of 64 stomach (or gizzard) stones was found beneath a sauropod shoulder blade in Howe Quarry. A specimen of another kind of sauropod (*Camarasaurus*) and two specimens of *Camptosaurus* were also collected.

There is a linguistic dimension to the name of *Diplodocus* as well. In 1899 the Carnegie Museum in Pittsburgh, with the interest, encouragement, and support of Andrew Carnegie, began to search for sauropod skeletons in the western United States. On 4 July of that year Carnegie collectors discovered a quarry northeast of Medicine Bow, Wyoming, which yielded the two major specimens combined in the *Diplodocus* mount in Pittsburgh. During the early part of this century casts of the skeletal mount were made and mounted in many European and Latin American museums, including the Museum d'histoire naturelle at the Jardin des plantes in Paris. Thus 'le diplodocus' entered the French language in a manner analogous to 'brontosaur' in English.

Barosaurus: length 23 metres, height at hips 4 metres, weight 40,000 kilograms. Several relatively complete specimens of this uncommon sauropod have been collected, but most of them have still not been described. It represents an extreme in a series of related dinosaurs (*Apatosaurus, Diplodocus, Barosaurus*) in a tendency for the neck to become elongated. According to quarry diagrams of the famous site at Dinosaur National Monument, near Jensen, Utah, the neck in large specimens of *Barosaurus* was probably 9 to 10 metres long. It was three times longer than the back, and, as in its Chinese contemporary *Mamenchisaurus*, the neck made up half the total length of the body. The increase was brought about by an increase in the length of each neck vertebra, for there are about as many vertebral segments in the neck as in *Apatosaurus* and *Diplodocus*. The tail vertebrae in *Barosaurus* are, however, shorter, and the base is not so powerfully constructed. Perhaps the whiplash at the end of the tail was also poorly developed.

Camarasaurus: length 17.5 metres, height at shoulders 3.8 metres, weight 31,000 kilograms (for the smallest known specimen: length 3.1 metres, height at shoulders 0.7 metre, weight 175 kilograms). *Camarasaurus* is a rather ordinary member of a stream of sauropod evolution separate from

that of the diplodocids. In camarasaurs, the large skulls are reminiscent of what might be imagined as a very odd kind of horse in which the head was as short as that of a bulldog. As noted before, the teeth are large and spoon-shaped, the forelimbs tended to be long, so that the back was either held parallel to the ground or sloped upward toward the shoulders, and relatively short tails did not end in a whiplash. The neck also became elongated in some members of this group but was usually held higher off the ground than in diplodocids.

Camarasaurus could be considered as a dinosaurian answer to an elephant. An elephant has a long trunk with which it strips foliage from bushes and carries it back to the mouth. The head is heavy because of the heavy jaws, large brain, and, frequently, tusks, and as a consequence the neck is short and powerful. *Camarasaurus* did not have a trunk, but it did possess a long neck. Because its jaws were weak and its brain was tiny, the head could be much smaller than that of an elephant. *Camarasaurus* could thus insert its meagre cranial equipment into bushes and trees on the apex of its 'trunk.' If there was some advantage to be gained from grinding the food, there is evidence that sauropods sometimes swallowed stones that probably accomplished the task in a muscular gizzard. Such being the case, *Camarasaurus* could be thought of as an elephant with its head at the tip of its trunk and its teeth in its stomach.

The most abundant sauropod in Morrison sediments, *Camarasaurus* probably did not compete with grazing diplodocids, for it was adapted to browse high off the ground on large bushes and trees. Because of the restriction of trees to areas along watercourses in semi-arid regions, the dinosaur may have been particularly abundant there. Half-grown specimens of *Camarasaurus* are found more frequently than those of other sauropods, suggesting that the young also

A Late Jurassic Plain

Como Ridge (Morrison Formation), near
Medicine Bow, Wyoming. Aurora Lake
(Lake Como) is in foreground of top photo.
Quarries scattered along escarpment have
proved rich in dinosaur skeletal material.

frequented watercourses where they were more exposed to drowning in flash floods than were the young of other sauropods.

Brachiosaurus: length 18 metres, height at shoulders 5.9 metres, weight 54,500 kilograms. Perhaps the most ponderous sauropod in the fauna, this highly modified camarasaur is known in North America from only a few incomplete specimens. The spines on the vertebrae in the chest region are unforked, long, and powerful. They provided support for a very long neck that could take the head to an elevation of more than 12 metres off the ground. The bones of the forelimbs, in spite of their huge size, were long and slender. The hindquarters, in contrast, were relatively short, giving the animal the profile of an enormous reptilian giraffe. *Brachiosaurus* is more completely known from Tanzanian specimens, which are also more abundantly found, but the African species is more slender and somewhat more giraffoid in its proportions.

Brachiosaurus may have been restricted to southern localities, in Utah and Colorado, where trees were more abundant. It could browse on branches far beyond the reach of other sauropods, but foliage at these levels had to be abundant enough so that the animal could find enough nourishment to balance its energy costs. Under most conditions *Brachiosaurus* would have been unable to find enough food to fuel a high metabolic rate such as modern mammals have. Its bulk alone would put it in peril from death through the accumulation of a heat load even if its metabolism were comparable to that of reptiles, as was almost certainly the case. The nostrils were large and may have provided a surface for evaporation to cool blood that in turn could be shunted to the brain, keeping it from being exposed to high body temperatures. During the day it may have sought shelter in the shade, and perhaps the organic fertilizer in the dung it left behind constituted a kind of biological compensation for the tree.

Small Carnivorous Bipeds (Theropods)

Ornitholestes: length 2.2 metres, height at hips 0.6 metre, weight 7 kilograms. The relatively complete skeleton of a small theropod was collected in 1900 by the American Museum of Natural History from an enormous excavation a few miles north of Medicine Bow, Wyoming. The excavation was called Bone Cabin Quarry after a cabin built by a shepherd from weathered blocks of dinosaur bone littering the surface of the ground. The quarry yielded fragments

Morrison Formation,
near Grand Junction, Colorado.
Brachiosaurus was first collected here.
Dakota sandstones cap hill.

of at least two dozen sauropod skeletons as well as of most of the other varieties of contemporaneous dinosaurs, although only the heavier skeletal pieces of large dinosaurs were usually found. Perhaps the site was once a shallow area in a broad stream, where the heavier parts of floating carcasses made contact with the stream bed and fell away. The small carnivore, however, was very lightly constructed, and its bones are nearly as thin as paper. In its bodily proportions, *Ornitholestes* was typical of many small raptorial dinosaurs. The eyes were large, the forelimbs were slender and prehensile, and the end of the tail was tightly bound together to form a rigid rod. The shins were too short for the animal to have been an exceptionally well adapted runner. *Ornitholestes* was not the smallest carnivorous dinosaur present on the Morrison plain. A tiny femur from the vicinity of Grand Junction, Colorado, was derived from an adult animal that probably stood 15 centimetres tall at the hips and weighed but 140 grams.

Coelurus: length 2.4 metres, height at hips 0.7 metre, weight 13 kilograms. The small carnivorous dinosaur *Coelurus* has provided a name for an entire complex of relatively small saurischian flesh-eaters (coelurosaurs) that lived at various times during the dinosaurian era. Unfortunately *Coelurus* has not yet been carefully studied, and available evidence suggests that it was a rather specialized form that may not have been representative of all the small dinosaurian carnivores. In the same way gerbils, jackrabbits, and wallabies may all look superficially alike, but they belong to separate mammalian orders. The neck of *Coelurus* was longer than in *Ornitholestes*, the tail was probably less rigid, and the legs were relatively much longer, indicating that the animal was more fleet.

Stokesosaurus: length ?4 metres, weight about 50 kilograms. Known from very fragmentary remains, *Stokesosaurus* was a short-faced carnivore with muscular hind limbs.

Elaphrosaurus: length 4 metres, height at hips 1.1 metres, weight 92 kilograms. A single well-preserved skeleton of this archaic ostrich dinosaur was collected several decades ago in Tanzania. The North American record is based on an arm bone, which together with the Tanzanian specimen provided the basis for the above dimensional estimates. The animal was probably a running omnivore like an ostrich. No cranial parts are known, but as the animal closely resembles its descendants in many details its jaws may too have been birdlike and the teeth reduced in size or absent. *Elaphrosaurus* was not as long-limbed as later forms.

Marshosaurus: length ?5 metres, weight about 100 kilograms. Very little is known of this carnivore, which evidently possessed a long skull and slender body.

Large Carnivorous Bipeds (Theropods)

Ceratosaurus: length 6.6 metres, height at hips 2.1 metres, weight 675 kilograms. *Ceratosaurus* was a medium-sized carnivorous dinosaur, which bore a triangular, bladelike horn above its flared nostrils. Peculiar fluted teeth collected in Tanzania indicate that the form was also present in east Africa. A single row of small bony plates extended from the ridge of the neck along the back to the base of the tail. The tail was very deep and compressed from side to side. Unlike most contemporaneous and later carnivorous dinosaurs, *Ceratosaurus* possessed four digits in the hand instead of three. Rugosities for muscle attachment to the limb bones suggest that the hind limbs were particularly powerful.

A famous dinosaur site was discovered near Canyon City, Colorado, in 1877. When the Marsh Quarry was opened, collectors were unable to focus their attention fully on their work because of the presence of hostile Indians nearby. Partly because of this distraction, and partly because suitable collecting techniques had not yet been devised, the first dinosaur specimens extracted from the strata suffered considerably in the process. The quarry later yielded many superb specimens, among them the first *Ceratosaurus* skeleton, collected during the seasons of 1883 and 1884, and now in the collections of the National Museum of Natural History in Washington, DC. A drawing of the skeleton was published in 1892, being the first reconstruction of a North American carnivorous dinosaur based on a single, relatively complete specimen. For this reason, and because of the characteristic horn on its nose, *Ceratosaurus* became a well-known dinosaur even though its remains are very uncommon. Only in recent years has another excellent *Ceratosaurus* skeleton been collected, near Grand Junction, Colorado.

Allosaurus: length 13 metres, height at hips 2.3 metres, weight 1700 kilograms (for the smallest known specimen: length 3.0 metres, height at hips 1.0 metre, weight 20 kilograms). *Allosaurus* differs most obviously from *Ceratosaurus* by its larger size, by the absence of a horn on the nose but presence of hornlike ridges of bone above and in front of each eye, by a tail that was smaller and more circular in cross-section, and by its heavily clawed, three-fingered hands. *Allosaurus* was the most abundant carnivore in the fauna, but few drawings are available of its skeleton as a whole. For many years the only skeletal mount in existence was

A Late Jurassic Plain

Morrison Formation, near Fruita:
channel sandstones. A *Ceratosaurus* skeleton was collected near here.

Sandstone boulders at Cleveland-Lloyd Quarry.
Weathered from Cretaceous Cedar Mountain Formation,
they overlie Morrison Formation.

in the American Museum of Natural History in New York, which was placed on display in 1908, crouched over a supine *Apatosaurus* skeleton. However, during the last 25 years some two dozen *Allosaurus* mounts have been exhibited all over the world, assembled from original materials extracted from a single site, the Cleveland-Lloyd Quarry, in central Utah. Discovered in 1927, the site was designated a natural landmark in 1967.

The Utah quarry represents the densest accumulation of dinosaur bone known from Morrison sediments (more than 10,000 have been collected in an average concentration of over 30 bones per square metre), and most of this material belongs to carnivorous dinosaurs (at least 44 individuals of *Allosaurus* and 6 other carnivore individuals are represented). Normally, only a few per cent of the dinosaur specimens collected pertain to *Allosaurus*, and there must have been some special circumstance that made this site a death-trap for carnivores. The following is one of several possible interpretations.

The trap originally consisted of a pond in an abandoned river meander that was surrounded by mud-flats and provided a dependable source of water during the dry season. Within a few weeks 5 sauropods, 5 camptosaurs, and 3 stegosaurs came to drink and were mired in or hampered in their movements by the mud. When they died, or were killed, they provided 120 metric tonnes of flesh, or enough food to sustain over 800 adult allosaurs for over a month. The carrion was intensely attractive to the carnivores, more so than the water had originally been to the plant-eating dinosaurs. The carnivores came, and being endowed with aggressive dispositions they not only tore the herbivore carcasses apart but killed each other as well (the escape of weaker individuals was hampered by mud). Predator car-

Cleveland-Lloyd Dinosaur Quarry: US Natural Landmark, Morrison Formation, south of Price, Utah. Movable shelters cover exposed dinosaur bones, mostly *Allosaurus*.

casses were attractive to smaller predators, and an ecologically unstable situation briefly arose whereby a substantial portion of the bait in the trap was provided by the flesh of predators themselves. Most of the *Allosaurus* specimens killed were subadults ranging between 3 and 12 metres in length and 20 to 1300 kilograms in weight. The skeletons of all the victims were dismembered and thoroughly mixed by trampling feet, with the bones of the original herbivorous dinosaurian victims at the bottom of the accumulation. Many bones were partly broken or splintered while they were still elastic. The supply of carrion dwindled, the waters spread, and a layer of freshwater limestone was deposited over a gently undulating surface of mud and bone.

A Site to Behold

Throughout the spring and summer of 1909 two collectors from the Carnegie Museum in Pittsburgh had explored northeastern Utah for dinosaur bones, but in vain. Nearing the point of despair, on 19 August they found broken pieces of fossil bone on a dry wash where it passed through the crest of a steeply dipping ridge of sandstone. The fracture surfaces on the bone fragments were probably fresh, and, as all collectors of dinosaurs do, they traced the fragments uphill toward their source. There on the crest of the ridge they found eight tail vertebrae of a gigantic sauropod imbedded in a great slab of sandstone. With picks and shovels they followed the string of tail vertebrae into the strata, and it led them to the greatest dinosaur quarry ever discovered in North America.

The dinosaur whose backbone they followed could well have been named 'Ariadnesaurus' but was actually a magnificent specimen of *Apatosaurus*. Since the time of its discovery the quarry has produced the greatest number of dinosaur bones, the greatest number of articulated skeletons of different kinds of dinosaurs, and the greatest number of sauropod skulls of any single site on the continent. It was made a national monument in 1915, at a time when many superb specimens were being removed from the quarry for storage and exhibition in the Carnegie Museum. A glass-walled display centre was constructed around the inclined layer of bone-bearing sandstone and opened to the public in 1958. Since then nearly three thousand additional bones have been exposed, while the public watches the ongoing search for new specimens and hears lectures about those that can be seen in the positions in which they were originally buried. The area is very rich in fossils and could produce excellent material from localities near the existing

Neck vertebrae and skull of *Camarasaurus*,
on quarry face, Dinosaur National Monument;
to the left: foot of a large sauropod

display centre. The wealth of Dinosaur National Monument remains to be fully exploited.

One hundred and fifty million years ago a broad stream dotted with sand and gravel bars flowed east by southeast through the monument. The river flowed throughout the year, and clams flourished in its tepid waters. Traces of roots and stems occur in the coarse sandstones, giving evidence of the invasion of sand-bars by woody plants and scouring rushes when water levels were low. Frogs, turtles, and small primitive crocodiles were present but not abundant. During the wet season, runoff entered the river system from the surrounding lowlands and water levels rose. Scouring occurred below areas of rapid flow, and adjacent banks and sand-bars were draped with mud. The carcasses of animals that had died and dried out on nearby plains floated on the rain-driven runoff. They were carried into the stream that flowed through the monument, where they lodged on submerged sand-bars. The lighter necks of the sauropods were often partly separated from the rest of the animal, but many carcasses were insufficiently decayed for them to be dismembered by currents. They were buried whole or in part in banks of sand and gravel beneath the sediment-charged waters. The skeletons were protected from predators and were able to enter the timeless vault of the fossil record.

The bones in the quarry do not represent a mass-death accumulation caused by a drought or a flash flood. The animals had died at different times, and skeletons in different stages of decay were swept up by the seasonal floods. Some skeletons were buried whole in the channel system, others dropped the heavier bones from the carcasses as they floated downstream, and many broke up completely, scattering bones the length of the channel. At least three high-water periods are represented in the sequence of bone-bearing sands in the quarry face, and the fossils accumulated over perhaps as much as a few decades. Because they represent natural mortality over an extended period, the dinosaur skeletons in the quarry give a clearer indication of the natural abundances of dinosaur populations living in the region.

All the major varieties of dinosaurs in the fauna are represented, and remains of the rare long-necked sauropod *Barosaurus* are unusually abundant. The most common animal in the quarry is *Camarasaurus*, and several superb skeletons have been excavated in varying stages of maturity. *Stegosaurus* is well represented, but the skeletal parts are usually more scattered than is the case with specimens of other dinosaurs, suggesting the animals died at a greater distance from the ancient stream channel. A few of the small bipedal dinosaurs, such as the herbivore *Othnielia* and carnivores *Coelurus*, *Stokesosaurus*, *Marshosaurus*, and *Elaphrosaurus* have not been recorded, possibly because they preferred less open environments. Neither is the primitive sauropod *Haplocanthosaurus* nor the giraffoid *Brachiosaurus*, probably for unknown ecological reasons. There is no question, however, that the monument is a superb cenotaph for an ancient time when giants walked the face of the planet.

The Largest of the Giants

The excavation of bones of the largest Morrison dinosaurs spans a century of collecting. The Cope Quarry, opened in 1877 in strata that were deposited late in Morrison time near Canyon City in southern Colorado, produced bones from several diplodocids. The bones closely resemble their counterparts in *Diplodocus* skeletons from other localities, where the largest specimens are estimated to have weighed but 16 metric tonnes. Two vertebrae and a thigh-bone were derived from one animal weighing about 23 metric tonnes, a very large shoulder-blade came from another animal weighing about 45 metric tonnes, and a gigantic vertebral fragment suggests the presence of a third weighing on the order of 100 metric tonnes. The animal represented by these bones has not certainly been identified, but the fact that three similar but very large *Diplodocus*-like specimens all came from the same locality suggests that they may collectively represent a distinctive but poorly known dinosaur. The largest known specimen of *Camarasaurus* and giant specimens of *Allosaurus* were also collected from this locality.

Located southwest of Delta in western Colorado, the Dry Mesa Quarry was discovered in 1971 and opened by paleontologists from Brigham Young University the following year. The quarry is situated in sediments deposited relatively late in Morrison time and has also produced bones of gigantic sauropods. A partial skeleton of a very large animal was excavated in 1972 and given the name *Supersaurus*. The animal may have resembled a large, long-necked diplodocid and weighed in the range of 45 to 80 metric tonnes. A huge shoulder girdle extracted in 1979 resembles the same structure in *Brachiosaurus* but belonged to an animal that, if it had the proportions of *Brachiosaurus*, weighed 90 metric tonnes and could lift its head nearly 15 metres above the surface of the ground. Appropriately, these shoulder bones serve as the basis for the name of *Ultrasaurus*. In the quarry on Dry Mesa were discovered the remains of *Torvosaurus*, a very muscular carnivore with a body length of

Dinosaur National Monument, near Jensen, Utah.
Scattered bones from a giant *Apatosaurus*
skeleton are visible in stream-channel
Morrison sandstones, exposed on quarry face.

about 10 metres and extremely powerful forelimbs. The skeletal parts of various dinosaurs in the quarry are largely disarticulated and range in size from a giant neck vertebra 1.3 metres long to a tiny tail vertebra 7 millimetres long. Delicate bones of flying reptiles and possibly also of primitive birds have been collected, as well as the more robust bones of small turtles and crocodiles.

The sauropod bones in these two quarries are among the largest known in the world. What factors limited the size to which these animals could grow, and how long did they live? Growth rings have been observed in sauropod bones which have been interpreted as evidence that the body temperatures of the animals fluctuated annually with the seasons and that they had low metabolic rates, like modern reptiles. A herbivorous reptile as large as *Ultrasaurus* would have had to consume about as much foliage per day as does a very large bull elephant. The implication is that, in practice, the size to which a land herbivore can grow is limited by the amount of time it can spend eating, not by the ability of bone, tendon, and muscle to resist the force of gravity. Counts of growth rings in bones of a sauropod from Madagascar suggest that the animal became reproductively mature, or two-thirds grown, at an age of 43 years. If it grew at a constant rate, then large animals would have died (ceased growing) at 65 years of age. If a slowdown in growth occurred, as was probable, greater ages would normally be attained.

It is a common observation that the larger a mammal is, the longer its life-span tends to be. Other factors are associated with a long life-span in mammalian species, such as the size of the brain. Data on the life-spans of zoo reptiles indicate that they live about as long as mammals would of the same body weight. The trend is not very well defined, for many zoo reptiles probably die as a result of health problems associated with improper surroundings, and others, such as turtles, enjoy very long average life-spans. If dinosaurs of different body weights followed the same general weight-versus-age trend as living reptiles, then the life-span of various dinosaurs could be estimated indirectly. The trend predicts that a dinosaur weighing about 1 metric tonne would live about 48 years, one weighing about 5 tonnes about 66 years, and one weighing about 25 tonnes about 89 years. An *Ultrasaurus* weighing 90 tonnes could expect to live about 114 years.

Big Game on the Plains

Because the Morrison fauna was so widespread and seems to have been so uniform geographically, it is likely that the dinosaur populations lived in equilibrium with their surroundings on the plain in a manner resembling a self-contained ecosystem. The nature of this ecosystem is unknown, because existing studies have for the most part been confined to the skeletal anatomy of the various different kinds of dinosaurs. Even in this area much work remains to be done. However, astronomers do speculate on the properties of quasars, which are so far removed in space that they lie near the limits of the observable universe. In a somewhat analogous manner, it might be useful to attempt to block out some of the attributes of the Morrison ecosystem, using a very narrow approach that is very poorly constrained by observational data. This approach may produce results that will be incorrectly interpreted as fact, but it may also provide a framework upon which academic exploration may proceed with greater precision.

The landscapes in which Morrison dinosaurs lived may be compared to landscapes in the Amboseli Basin in southern Kenya, where the climates are hot and monsoonal, and there is a very pronounced dry season. Lake Amboseli covers much of the basin, but its flat bed remains dry except after heavy rains. The vegetation is supported largely by subsurface water from nearby Mount Kilimanjaro in Tanzania. Wildlife and domestic herds migrate across the basin, leaving behind skeletons of animals that die there. If one counts the number of skeletons seen in the basin and compares the result with the animal populations inhabiting it on an annual basis, it is apparent that skeletons of the smaller animals are more often destroyed than those of the larger animals. Even when they are preserved, they are more difficult to see. However, a simple mathematical relationship can be deduced to compensate for the anomalously low number of smaller skeletons and produce an estimate of the original populations.

The relative abundance of different kinds of dinosaur skeletons that have been collected from Morrison sediments can be calculated from data in paleontological publications and museum field records by estimating the minimum number of individuals needed to account for all the skeletal parts collected (Table 1). Specimens of *Allosaurus* at one site (Cleveland-Lloyd Quarry) and *Diplodocus* in another (Howe Quarry) have been excluded from the counts, for they are unusual occurrences. The relative abundance of dinosaurs of different adult weights that once lived on the Morrison plain can then be estimated by using the

Amboseli formula. For example, the results suggest that although the giant sauropod *Camarasaurus* is the most common dinosaur in collections, the tiny ornithopod *Nanosaurus* is calculated to have been numerically the most common dinosaur on the ancient plain.

TABLE 1   Relative Abundance of Dinosaur Skeletons in Morrison Sediments

% INDIVIDUALS COLLECTED		% INDIVIDUALS PRESENT (ESTIMATED)	
Camarasaurus	20.6	Nanosaurus	13.1
Apatosaurus	16.1	Dryosaurus	12.0
Diplodocus	14.1	Othnielia	11.7
Stegosaurus	13.0	Allosaurus	10.0
Allosaurus	11.6	Coelurus	9.3
Camptosaurus	8.1	Camptosaurus	8.8
Dryosaurus	3.9	Stegosaurus	6.9
Barosaurus	2.0	Ornitholestes	6.1
Ceratosaurus	1.9	Camarasaurus	4.8
Othnielia	1.8	Diplodocus	4.4
Haplocanthosaurus	1.7	Apatosaurus	3.3
Coelurus	1.2	Stokesosaurus	2.5
Brachiosaurus	1.0	Ceratosaurus	2.5
Ornitholestes	0.6	Marshosaurus	1.9
Marshosaurus	0.6	Elaphrosaurus	1.0
Stokesosaurus	0.6	Haplocanthosaurus	0.8
Nanosaurus	0.5	Barosaurus	0.4
Elaphrosaurus	0.3	Unnamed ornithischian	0.2
Unnamed ornithischian	0.2	Brachiosaurus	0.2
Unnamed sauropod	0.2	Unnamed sauropod	0.1

Estimates of the relative abundance of different kinds of dinosaurs, together with the adult body weights for each variety, can provide the basis for some ecological extrapolations. Suppose, as seems likely, that the dinosaurian herbivores of the Morrison plain were metabolically similar to reptiles. The proportion of the total amount of vegetation available that was required to feed each population of dinosaurs can then be estimated. Thus *Camarasaurus* herds on the Morrison plain would have eaten about three thousand times as much plant food per day as all of the *Nanosaurus* flocks combined, given that the plant foods eaten by both kinds of dinosaurs were equally nourishing. The population density of Morrison dinosaurs can also be estimated. Suppose that the plants on the Morrison and Amboseli plains produced a comparable amount of food and that this food

were harvested by the animals with comparable efficiency. Each square kilometre in the Amboseli basin supports, on average, about 25 mammalian herbivores (excluding rodents and other small forms). A series of calculations, based on postulated abundances, body weights, and food requirements, interpreted according to the Amboseli analogue in food production and cropping efficiency, leads to the result that the same area on the Morrison plain would have supported an average of 13 herbivorous dinosaurs.

These numbers imply profound differences in the appearances of big game herds on the Morrison and Amboseli plains. Because, on average, each dinosaur would have weighed 7000 kilograms (the weight of two elephants combined) and each mammal, on average, weighs 200 kilograms (the weight of a zebra), the vegetation would have been actually feeding nearly twenty times more herbivorous dinosaur flesh than Amboseli herbivorous mammal flesh per square kilometre. Low reptilian metabolic rates in combination with low metabolic rates associated with animals of great size would permit a huge 'standing crop' of giant reptiles. Further, big animals are easy to see. Amboseli supports only one elephant per 4 square kilometres, but the Morrison plain is calculated to have supported 40 times as many animals that were twice as heavy as an average elephant. The result implies a visual impression of stupendous herds of colossal animals. Such a vista would seem as if it belonged on another planet to a wildlife biologist accustomed to east Africa, and utterly unbelievable to a typical North American sportsman.

The 'standing crop' of large carnivorous dinosaurs was, relative to that of the herbivores, even greater than in Amboseli. Following the line of reasoning applied to the herbivores, there would have been two specimens of the great bipedal carnivore *Allosaurus* on each square kilometre of the Morrison plain, along with many other smaller varieties of flesh-eaters. The total amount of flesh produced by herbivorous dinosaurs each year, which could be eaten without reducing the size of the game herds, could comfortably feed all the flesh-eaters, *if* they too had reptilian metabolic rates. If they possessed mammalian metabolic rates appropriate for animals of their respective sizes, only a quarter of the required food could have been provided without jeopardizing the survival of the game herds and rapidly producing famine among the carnivores. According to these figures, the dinosaurs were only as active as reptiles, and a tourist on safari on the Morrison plain could be counselled to apply only about a quarter as much caution to each *Allosaurus* that he met as he would to each lion encountered on the Amboseli plain!

There would also be homely (and therefore attractive?) baby dinosaurs. Morrison sediments are unusual in the wide range of growth stages represented by well-preserved specimens (Table 2). The ratio of juveniles to adults, after using the Amboseli formula to offset preservational bias against small skeletons, suggests that mortality was somewhat higher among infant animals, which would not be surprising. Although the range is considerable (162 grams to 258 kilograms), the average weight represented by baby dinosaur specimens would about equal that of very large men. However, these babies were much smaller relative to their parents (about 1/45th the weight) than human babies are relative to their fathers (about 1/25th the weight). Because of their small size and presumed inability to travel far, it is likely that most dinosaur babies (with the possible exception of the sauropods *Haplocanthosaurus* and *Brachiosaurus*, where no juveniles have been found) hatched on the plain.

TABLE 2 Growth Stages Represented in Morrison Sediment

	SMALLEST BABY (KG)	LARGEST ADULT (KG)	RATIO (%)
Othnielia	0.162	18.8	0.9
Dryosaurus	4	100	4.0
Allosaurus	20	1700	1.2
Camptosaurus	76	1000	7.6
Diplodocus	118	15,900	0.7
Stegosaurus	163	6400	2.6
Camarasaurus	175	31,200	0.6
Apatosaurus	258	42,600	0.6
Average	102	12,365	2.27
Human	3.5	90	3.89

Hatchlings of the biggest dinosaurs must initially have weighed no more than about 10 kilograms. This weight limit was because the eggshell had to remain porous, so that air could reach the embryo, and yet be strong enough to reduce the hazard of premature breaking. Eggs much heavier than this would suffocate the embryo because the shells would have to be too thick in order to support the additional weight. Because baby mammals depend on their mothers' milk for food, they indirectly depend on the same food that their mothers eat. Infant herbivorous reptiles must depend on nutritious food other than milk, and in most cases they eat insects and other soft-bodied animals. One must not have been too shocked if young sauropods were rather indiscriminate in what they ate. Perhaps they even devoured the eggshells after they hatched, for no shells have yet been found in Morrison strata. Like all babies, they must have needed calcium to grow.

Morrison dinosaurs differed from modern east African mammals in many ways. Because of their generally smaller brain size, the behaviour of dinosaurs would probably have seemed like that of an unimaginative specialist in comparison to the more flexible 'jack of all trades' approach of many living mammals. But what they did they did well. Few would have questioned the aptitudes they did possess, just as few question the proficiency of Nile crocodiles (which have brains comparable in size to that of an average Jurassic dinosaur) at outperforming human swimmers. Nevertheless there must have been an upper limit to the number of different behaviour patterns that a typical dinosaur could carry out, and because of this their skeletons probably indicate more precisely what these few things were than do those of mammals. Consider the behavioural demands an otter makes on its forefeet, compared to those made by a crocodile.

Mammals characteristically depend a great deal on scent in taking the measure of their world, but dinosaurs seem to have relied more on vision. Their eyes were on average one and one-half times the diameter of the eye in living mammals. Unlike modern mammalian herbivores, and many later plant-eating dinosaurs as well, Morrison dinosaurs generally lacked powerful jaw muscles and grinding dentitions. They were evidently gulpers, not chewers of food. Because the heads of the plant-eaters lacked both large jaw muscles and large brains, they would have appeared incongruously small, consisting of all mouth and eyes.

The Morrison dinosaurs cropped foliage over a much broader vertical range than do the mammals of Amboseli. They were not so restricted to feeding on herbaceous vegetation growing close to the ground. This was probably due to an absence of grasslike prairie plants in semi-arid regions at that time rather than to any special feeding strategy of plant-eating dinosaurs. Modern grasses may be considered as specialized plants that retain their centres of growth in the soil in order to survive severe browsing. Dinosaurian herbivores probably inflicted heavy damage on shrubs and trees, and this damage must have been very obvious in Morrison landscapes. Some plants bore very large thorns, and others protected their centres of growth with heavy scales. Because its denizens nearly equalled trees in stature, the Morrison plain was effectively a two-dimensional topography in which giant animals sought appropriate patches in which to feed. They could escape predators only

Elephant browsing in high brush, Luangwa Reserve

Modern open forest of primitive aspect, New Caledonia

Waterfall (chute de la Madeleine), in bushlands
of primitive aspect, New Caledonia

by flight, but, in contrast to the relatively small, energetic, and intelligent Amboseli herbivores, they probably more often stood their ground. If the animals were weak they very likely were killed and eaten on the spot, rather than pursued and brought down after a chase.

Sauropods' feet resembled elephants', and, like elephants, they probably preferred to browse on firm ground. There are no obvious anatomical indications that any of the large herbivorous dinosaurs were as amphibious as a hippo. Quarry evidence suggests that sauropods preferred the company of their own kind to that of other dinosaurs. The plain might thus have been dotted with groups of diplodocids, each composed of one variety, feeding in fern fields. It would have been fascinating to see several specimens of *Diplodocus* with their heads turned to the side, slashing their bullwhip-like tails at an *Allosaurus* that came too close. Because *Apatosaurus* was a larger animal, it may have fed in richer sites, but exactly how its habitat differed from that of *Diplodocus* remains unknown. Clumps of *Camarasaurus* spread havoc and destruction among the smaller trees, and *Dryosaurus* individuals jostled each other to glean branches that fell to the ground. Areas where the vegetation was sparser and dominated by cycad-like plants were the home of the stegosaurs and smaller bipedal ornithopods. Their narrow beaks indicate that these animals were more discriminating in their food preferences than were the bulk-feeding sauropods.

The pattern of dinosaur groups was repeated over and over across the broad Morrison plain, and the giant herds probably migrated in response to seasonal changes in rainfall. Indeed, the food requirements of such large animals necessitated more or less continuous movement. There were 'islands' in the plain where environmental conditions were different. Some areas, possibly such as windfalls or logjams, offered abundant shelter for tiny animals that was not normally available. A few regions were covered with lush savannah growth and an unusual number of trees. These seem to have sheltered huge, giraffe-like brachiosaurs and giant diplodocids. Perhaps exceptional environments became more frequent near the southern edge of the plain. On the whole the fauna seems to have been uniform across the vast expanse of semi-arid interior flatlands.

The seasons came and went, and seemingly everything changed and nothing changed. There were images of sauropod necks and backs glistening in monsoonal rains. Of lush savannahs, clouds of insects, and streams of migrating giants. Of great, broad swaths of red and black mud cut by game trails. There were images of mechanical, emotionless reproduction, of surpassing stupidity, of masses of fibrous

dung, of blood-soaked earth, and of chunky hatchlings arriving with the rains, representing the hope for generations and generations to come – the future we now inhabit was astrophysically remote. A human would have been ignored, unless viewed by a *Ceratosaurus* with an ache in its entrails. The attack would have been short, vigorous, and effective.

The rains ceased, and there was another set of images. Of ferns withered into a brown woolly carpet extending to the horizon, and dust and particles of dinosaur dung blowing across dry lake beds. Of giant bones bleaching in the sickening heat of midday, and of a solitary, starving stegosaur motionless in the sparse shade of a tall cycad. Of fleet lizards, tiny running crocodiles, and minuscule scampering dinosaurs. The herds had departed; it would be safe for us for a while. Far away on the western horizon the mountains loomed blue-green in the distance. Snow never fell on them.

Elsewhere: An Atlantic Mediterranean

Beyond the eastern limit of the plain, semi-arid lowlands and eroded rift mountains extended to the edge of a sea in the 'centre of the earth,' a sea that covered the growing oceanic rift between North America and Europe. The Grand Banks of Newfoundland were still in contact with Spain, but barrier reefs grew along the western shore of a narrow ocean off Nova Scotia and New England. Brazil touched the Guinea Coast, so that the dinosaurs of the Morrison plain could mingle with their counterparts in the famous Tendaguru fauna of east Africa via a southern land route.

Europe was an archipelago scattered through a shallow, tropical sea. In southern Germany a broad lagoon separated a barrier reef from islands close by to the north. The floor of the lagoon was covered with a thick mantle of chalky mud, lying among low submarine hills formed from dead sponge reefs. Few animals lived in the ooze, for the bottom waters were deficient in oxygen. When heavy storms passed by, the muds were disturbed, and undercurrents carried blizzards of chalk to basins lying between the sponge hills. These basins were about 150 metres deep and extended for several kilometres. Here the chalks were compressed into fine-grained limestones, which have been extensively quarried for lithography (Solnhofen Limestone, 200 metres thick).

Fossils are not common in the lithographic limestones, but over the years many have been collected and the quality of preservation is outstanding. Nearly all the fossils are of marine organisms, such as free-swimming sea lilies (crinoids), coiled cousins of the chambered nautilus (ammonites), and 7-centimetre-long, heavily scaled fish (*Leptolepis*). There are a few fronds of land plants, which were probably blown from the islands during storms. These include foliage of ferns, cycads, ginkgos, araucarias, and cypress-like conifers that may have been tolerant of arid or saline environments. Skeletons of several different kinds of small pterosaurs have been exquisitely preserved, but these are very rare. Land animals are represented by unique specimens of a 60-centimetre-long carnivorous dinosaur (*Compsognathus*), a lizard-like animal related to the New Zealand tuatara, and an infant crocodile.

The fame of the lithographic limestones of southern Germany rests on five additional skeletons and skeletal fragments of small terrestrial vertebrates. All belong to one species, and if they had been discovered in sediments deposited on the Morrison plain they would have been considered as pertaining to small, rather uninteresting carnivorous dinosaurs. Their generally unspecialized skeletons would have been recognized by their long arms, which were supported by large collar bones. These bones were fused together into a 'Y'-shaped structure where they met in the centre of the chest – a wishbone. However, abundant feather impressions are preserved with the skeletons in the fine-grained limestone that would not have been preserved in the Morrison muds. They reveal that this little 'dinosaur' could run, flap its feather-lined arms, and fly. The recurved claws on its hands indicate that the animal was a raptor, in spite of the fact that its beak was not hooked but rather bore teeth within an ordinary dinosaurian muzzle. The bony axis of the tail was very long, but it supported a double row of feathers. *Archaeopteryx* was certainly a very unusual dinosaur, and it was certainly a very atypical bird.

Microvenator
and
Tenontosaurus

EARLY CRETACEOUS,
ABOUT 110 MILLION YEARS AGO

In northern Wyoming, a small
ornithopod (*Tenontosaurus*)
devours fruit from a cycad-like
(benettitalean) plant;
an egg-stealing theropod
(*Microvenator*) watches.
Background trees are
primitive conifers.

The Early Cretaceous

The Wealden of England

At the beginning of the Cretaceous Period the record of dinosaurs in North America was interrupted for an interval of about 25 million years, between 144 and 119 million years ago. During this time the interior of the continent stood, as Africa does today, well above sea level, and there were few interior floodplains and lake basins where sediments, and with them the bones of dinosaurs, could be preserved. However, a land-bridge linked Labrador and southern Greenland with the British Isles. South of the bridge, at its broadest the growing Atlantic Ocean was still less than 1500 kilometres wide. There is a record of dinosaurs in Britain and western Europe, and closely related animals probably inhabited climatically similar areas of North America.

The dampness of the British climate today is proverbial, and the English countryside was once covered with a nearly continuous carpet of forest. Weald is an old English word for forest that came to be applied specifically to the forested region south of London. Plant fossils can be found in Cretaceous strata there, and microscopic but identifiable pollen and spores often occur in great numbers in a handful of sediment taken from a road cut. Dinosaur bones and skeletal fragments are, in contrast, much less abundantly preserved and can very rarely be found in small exposures of fossil-bearing sediments. That many specimens have been collected from strata of early Cretaceous age in the Weald of southern England is a great tribute to the interest and enthusiasm of English naturalists. If the region were a barren desert, the abundance of dinosaur remains in its badlands would make it one of the richest fossil fields in the world.

England was then near the southern edge of the north-ern temperate zone and had a climate like that of the Mediterranean coast of Africa today. It was a land of bays, block mountains, and earthquakes. Great sheets of sediment (Wealden Group, 900 metres thick) eroded from the mountains onto outwash plains and mud-flats during a brief season of storms. A long dry season followed when stream channels emptied and ferns withered and yellowed on the lowlands. Termites harvested dry wood lying on the ground. Then thunderstorms carrying the returning rains brought lightning with them, and wind-whipped fires swept through the prairies of dried ferns. Conifer saplings were killed by the fires, but the ferns, like modern grasses, sprouted form stems that lay protected in the soil. Torrential downpours left jack-straw piles of logs decaying along delta streams, and mud-flats turned into vast verdant carpets of tall, wavering fern. There were lacewings, dragonflies, crickets, weevils, beetles, and roaches, but higher social insects, such as ants and bees, had not yet appeared on earth, and the beauty of butterflies was still to come.

Spiny-leaved conifers grew in sheltered vales, and bizarre, parasol-crowned ferns (*Weichselia*) spread on arching roots across the open hillsides. Lines of cycads and cedar-like trees followed stream courses across the fern prairies to shallow, salty lagoons. Freshwater marshes near the sea supported scouring rushes 2 metres tall and peculiar many-headed tree ferns (*Tempskya*). Cedar-like conifers (cheirolepidiaceans) grew in drier lowlands near the bays, and branches carrying their fat, succulent leaflets were often blown into the shallow bays in storms.

A strongly seasonal climate and vegetation dominated by fern prairies, with shrubs and trees limited to stream courses, recall the vast late Jurassic Morrison plain in the western interior of North America. Yet brontosaurs were not the dominant animals in the early Cretaceous of England. There are a few bones of whip-tailed diplodocids and fragmentary skeletons of animals that resembled brachiosaurid giraffe dinosaurs (*Pelorosaurus*). A third sauropod (*Titanosaurus*) belonged to a group that came to dominate Southern Hemisphere faunas later in Cretaceous time and that can easily be identified by the vertebrae of its tail. The vertebral bodies are strongly cupped in front and bear a large hemispherical projection in back that fits into the cup of the following vertebra.

Skeletal parts are seldom found of squat plant-eating dinosaurs (nodosaurids, *Hylaeosaurus* and *Polacanthus*) which bore bulky spines along the flanks of their shoulders and carried a heavy armour of bony spines on their backs. By far the most abundant dinosaurian remains are those of ornithopods. These herbivorous dinosaurs were fundamentally bipedal walkers and runners that depended on movement rather than armour for defence. The Wealden ornithopods were rather similar to those of the Morrison plain in both the number of varieties and the range of sizes represented in the fauna. However, the largest ornithopods were much more common. It is difficult to understand why this was so in view of the similarities between the late Jurassic and early Cretaceous environments and vegetation. Perhaps in some way the large Wealden ornithopods outcompeted the giant, bulk-feeding sauropods. The fauna was dominated by ornithopods as soon as the Wealden record began, and these animals continued, apparently without great change in abundance, throughout the 25-million-year span of Wealden sedimentation.

One of the most completely known small ornithopods is *Hypsilophodon*, 20 skeletons of which were collected from a single fossil stream bed. Evidently a herd of these animals, including both half-grown and adult specimens, had been overwhelmed in a flash flood. When fully grown the animal measured about 2.3 metres in length, stood 60 centimetres high at the hips, and weighed about 15 kilograms. It was very fleet and possessed long, slender hind limbs. The beak was a short and deep nipping device, and the teeth were suited for cutting, not chewing food. The animals were primitive in that they possessed teeth in the front of their beak and had no more than a dozen or so cheek teeth in each jaw. Only six vertebrae braced the hips against the backbone, and a well-developed inner toe was retained in the foot. A second, somewhat larger running ornithopod (*Valdosaurus*) is known only from a very few isolated bones. The animal probably measured between 3 and 4 metres in length and weighed approximately 100 kg.

The large ornithopods are typified by *Iguanodon*. These often ponderous creatures were advanced over their smaller relatives in that the number of functional teeth had doubled in each jaw, and replacement teeth supported those in use from below through a number of vertical ridges and grooves. Mammals such as horses and elephants grind their food on a horizontal plane between the upper and lower teeth. However, these dinosaurs ground plant fibres in a vertical plane through a system of pressure links between the bones of their cheeks and the lower jaws. In order to support their more massive bodies, iguanodonts buttressed the bones of their hips against as many as eight vertebrae. They characteristically possessed a strange spinelike thumb, and the inner toe of the foot was either reduced or absent. They left a three-toed track in soft ground.

One can visualize a herd of one of the most famous of all dinosaurs (*Iguanodon bernissartensis*) clustered in the shade of flat-topped conifers rooted in a dry stream bed. Most are quiet, but a few are slowly moving about, to the discomfort of their neighbours, who would otherwise be contented in their dusty beds. One large old cow, measuring 8 metres in length and weighing nearly 4 metric tonnes, slowly lifts her slender forequarters off the ground and extends her mooselike head into the foliage surrounding the lowest branches. Falling slowly back with a mouthful of fronds, she commences to chew thoughtlessly. The bones of her cheeks first bulge outward, and the lower margins of the jaws come together as they close. The cheeks then collapse, and the lower jaws spread as they descend before the next deliberate, grinding chop. Sometimes the animals would hook their forepaws over branches to hold them as they stripped the leaves. The heavy, hornlike cone of the thumb could be swung like a mace in self-defence. But when crossing open ground the big iguanodonts walk on both their forelimbs and hind limbs, recalling the quasi-comic, quasi-majestic gait of llamas.

One of the most unusual dinosaurian occurrences in the world lies deep within a Belgian coal-field. During early Cretaceous time, low limestone hills containing seams of coal were attacked by acid ground waters descending through fissures in the rocks. Subterranean caverns were formed, and the roofs of some of these collapsed so that they were open to the air. One such pit, some 60 metres wide and over 180 metres deep, was at least partly filled with water. More than two dozen large *Iguanodon* specimens entered or fell into it and died there over an undefined span of time. Their skeletons were preserved in in-fallen sediments, among blocks from the walls of the pit, together with those of many different varieties of fish, turtles, and small crocodiles. Ground water cannot erode limestone below sea level, so the land surface which the iguanodonts and parasol ferns lived must have been about 200 metres above sea level. Coal miners found *Iguanodon* skeletons in two shafts separated by a vertical interval of 34 metres in the pit, and it is highly likely that many more skeletons remain undiscovered. All the skeletons collected were of animals that were over three-fourths grown, and most were fully adult. Was this because juveniles were able to escape, because only adults lived in the low hills nearby, or because only a single herd of adult animals accidentally fell into the pit?

There was a smaller species of iguanodont in the English fauna named *Iguanodon mantelli*, which measured a little over 5 metres in length and weighed only 1.5 metric tonnes. The smaller animals had shorter and more slender forelimbs, the thumb-spine was less well developed, and they held their forelimbs clear of the ground as they walked. Their ecological niche is unknown; perhaps because of their smaller size they were creatures of the forest.

The teeth of large carnivorous dinosaurs are commonly found, but skeletal fragments are very rare. One large theropod (*Altispinax*) had long vertebral spines that evidently supported a narrow sail-like structure along its back. A few bones are known of a small raptor-like dinosaur (*Aristosuchus*). A recent discovery can be taken as symbolizing how much of the dinosaurian world remains unknown. Although Wealden strata in England have yielded bones and skeletal fragments of dinosaurs for more than a century, none had been found to indicate the existence of one peculiar, moderately large carnivore. Then a giant, 30-centimetre-long claw was found, signalling the location of a scattered skeleton of *Baryonyx*. The skeleton has not yet been reconstructed, but what has been described is indicative of a very peculiar creature. The muzzle is long and crocodile-like, and the jaws are lined with an unusually large number of teeth. Unlike in crocodiles, the back of the skull is deep. The neck was rather long, and the vertebrae imply that it was normally held in an extended position, instead of being bowed upward, as it is in most carnivorous dinosaurs. The forelimbs were quite large, and the hind limbs were probably rather short, which would have given the animal a rather crocodiloid appearance. A carnivorous dinosaur with such an atypical skeleton must have had atypical habits. Jaw fragments collected in the Sahara indicate that *Baryonyx* also lived in Africa.

Sediments of Wealden age in western Europe contain fragmentary remains of the forerunners of animals that became common later in Cretaceous time. In southern England was found the top of a skull of a small ornithischian dinosaur named *Yaverlandia*. The structure was thickened in a manner of suggestive of later bone-headed dinosaurs (pachycephalosaurids). The posterior half of a small herbivore skeleton, *Stenopelix*, was collected in Germany and contains a pelvis resembling those of horned dinosaurs (ceratopsia). Fossil feathers have been collected in Spain, and bones of ancestral oceanic diving birds occur in England (hesperornithiforms, *Enaliornis*). Far to the north of Norway, on the island of Spitsbergen, *Iguanodon* trackways have been found that are identical to those in southern England. Footprints of carnivorous dinosaurs, the scourge of the iguanodonts, also occur there. The island is now considered to be part of Europe, but when the iguanodons were alive the Arctic Ocean basin was only beginning to open. Spitsbergen then lay north of Greenland, near the Canadian arctic islands.

The Great Green North

At the beginning of Cretaceous time, environments in the polar regions of the Earth differed from those closer to the equator, but they were also very different from environments in polar regions today. Now, the greatest threat to polar life is cold. The summer midnight sun and winter midday moon are relatively minor marvels compared to the terrible seasonal changes in temperature. Then, the polar regions were warm, and differences in illumination between the pole and the equator were biologically much more important than at present. The fossil remains of forests that grew within the Antarctic Circle of that time indicate that many of the trees did not shed their leaves during the period of winter darkness. Stump fields show that the trees were often small, the base of their stems measuring between 8 and 22 centimetres in diameter, and were separated from each other by an average distance of only 4 metres. Seasonal growth rings in the woods are clearly marked, and areas where growth was locally accelerated (reaction wood) indicate that almost half of the trunks were tilted. Ground-dwelling plants were abundant. Presumably the same adaptive characteristics were typical of forests living within the Arctic Circle as well.

In order to account for the warm polar temperatures, it has been proposed that the Earth's axis of rotation was more nearly vertical, so that the polar regions would be more evenly illuminated and warmed throughout the year. However, if this were true, the Sun would always be too near the horizon to produce the annual growth flush that is clearly marked in the fossil stumps. It is more likely that the polar regions were warmed by the greenhouse effect of higher concentrations of carbon dioxide in the atmosphere

Isachsen Formation: coal-bearing strata.
South of Sand Bay, Axel Heiberg Island,
Northwest Territories

and that the annual flush was produced by a seasonally high midday sun and 24-hour daylight. Even so, the Sun could never have risen higher than about halfway to the zenith, and most of the illumination was at a very low angle even in midsummer. The abundance of reaction wood suggests that trees were leaning toward the light, just as house plants do on a window sill.

Trees that were as closely spaced as those in the antarctic stump field would have shaded each other, stunting each other's growth because the Sun was usually so low. This fact presents an interesting paradox. In tropical regions, plants compete with each other by shading and attempt to surpass their neighbours for the resource represented by sunlight. Thus, if rainfall is sufficient, grass will be shaded out by bushes and bushes will be shaded out by trees. Tropical trees tend to be flat-topped in order to intercept as much of the intense, nearly vertical sunlight as possible. Trees in the subarctic taiga are conical in shape in order to intercept sunlight that is usually at a lower angle in higher latitudes. A deep carpet of shade-loving mosses and bushes grows on the forest floor. Forests can no longer grow far above the Arctic Circle because it is now much too cold there. However, the struggle for nearly horizontal streams of sunlight must have been intense between arctic trees of early Cretaceous time, and the competition would have been in a

horizontal rather than vertical direction. Saplings would have flourished at the outer edge of a clump of trees where the light was strong, and trees at the centre would have died because of shading. The ancient arctic forests may have grown in expanding rings, leaving a field of closely spaced stumps in a carpet of moss and shade-tolerant bushes with the centre of the ring.

Dinosaurs inhabited these polar forests, for footprints of *Iguanodon*-like herbivores occur in northern Alaska, very near the location of the North Pole during early Cretaceous time. What special adaptations the animals possessed for living in boreal regions is unknown, for no skeletal parts have yet been found. Were the contours of their bodies more rounded in the absence of the midday heat stress of the tropics? Did they accumulate deposits of fat in their tails or beneath their skins in order to survive the periods of winter darkness, or did they migrate instead to more southerly latitudes?

During the last half of the 25-million-year gap in the history of dinosaurian evolution in North America, or between approximately 132 and 119 million years ago, rifting began under the shallow arctic sea and a deep-water Arctic Ocean trench began to form. Major uplifts occurred simultaneously in the lands now bordering the Arctic Circle. A great sheet of sediment spread over the Canadian

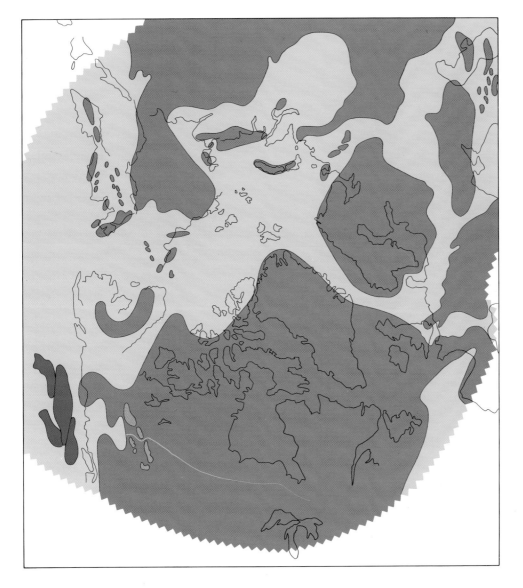

135 million years ago

Northern polar region.
Large Pacific islands
are approaching North America;
northwestern Europe remains in
contact to the east. River
systems flow off northeastern
highlands into northern seas.

arctic islands (Isachsen Formation, measuring 1.5 kilometres in thickness in Canada where it is nearest to Greenland, and thinning to about 100 metres near the edge of the Arctic Ocean in the west). Great rivers 20 metres deep carried sand and plant debris to the north, across an enormous coastal plain dotted with lakes, swamps, and arcing streams. Under the mild arctic climate a luxuriant blanket of cone-bearing trees followed the plain down from the southern highlands. Additional varieties of trees blended into the patchwork of forest rings which continued across the broad lowlands to a shallow northern sea. Cycad-like plants were less common, preferring relatively warm, well-drained sites. Many different kinds of ferns grew on sandbars and riverbanks. Others sprouted within dank moss moors and hung like garlands from the crowns of bleached,

decaying trunks in shaded areas behind the advancing forest fronts. Although the spoor of dinosaurs has been found in the fossil muds of Spitsbergen to the east and Alaska to the west, it has not yet come to light in the soft, warm forests of this archaic boreal Eden. It will be found.

Far to the south, a long, low interior plain extended from the arctic gulf through the Mackenzie Valley into what is now the eastern Rocky Mountains in the United States. West of the plain were a series of mountainous islands and peninsulas, separated by narrow channels and embayments of the Pacific Ocean. These terrains were in the process of slowly consolidating. The inland plain would soon fall beneath sea level. For the remainder of Cretaceous time this inland seaway would repeatedly fill with sediment from eroding highlands, to founder again before rising oceans.

When the northern sea occupied northwestern Canada 110 million years ago, a large river system emptied into it from the east. This system drained the region of Ontario south of Hudson Bay, where it filled a shallow basin with sediment (Mattagami Formation, 100 metres thick). Plant growth vigorous enough to produce seams of low-grade coal and the accumulation of high-quality ceramic clays indicate warm climates with heavy, seasonal rains. Plant microfossils bear witness to a profusion of ferns and forests of cone-bearing trees. Unfortunately the acid soils evidently caused bones to disintegrate. Warm, wet climates also left ceramic clays on the east coast of Nova Scotia, and rivers emptying into the Atlantic made deltas now buried offshore on the eastern Canadian continental shelf. During early Cretaceous time all of Canada was verdant, and there was an abundance of food for dinosaurs. So far, however, no dinosaur bones have been found.

Dinosaurs in Motion

Dinosaurs tracks, however, are abundantly preserved in the canyon of the Peace River in northern British Columbia. During the time of the Wealden dinosaurs, a gulf from the Pacific, resembling the modern Gulf of St Lawrence, reached the interior plains of western North America. Its fiord-like passage was destroyed as large and rugged island land masses from the Pacific, soon to become the Rocky Mountains, crumpled against the edge of the continent. Then, a few million years after the Wealden record came to a close in England (119 million years ago), the shallow northern seaway spread southward toward the Peace River area. There it ended in marshes and swamps, beyond which semi-arid plains extended far to the south and east. Across central and southern Alberta, these plains were dotted with hills of more ancient strata. The hills were buried slowly by broad fans of sediment, spreading from the region of contact of the rugged island land mass in northern British Columbia. This was where the Peace River dinosaurs crossed sand flats and bayous, which were drained and flooded daily by the tides, along the northern edge of one of the fans as it merged into swamps by the sea (Gething Formation, 500 metres thick).

Fossil trackways are special, for they provide petrified images of dinosaurs in motion. The most common Peace River tracks were made not by *Iguanodon* but by another bipedal plant-eating dinosaur (*Amblydactylus*), which had broader, three-toed feet. Its tracks are indistinguishable from those of duck-billed dinosaurs which were so common in North America 40 million years later. Little is known of the animals except that they weighed about 2 metric tonnes, their bodies swung from side to side as they walked, and they left no tail prints behind. Trackways of baby animals occur together in small groups, and adult trackways are accompanied only by those of half-grown or older animals. Juveniles touched their forefeet to the ground much less frequently than did the adults. In one trackway series, 11 animals were walking in a broad front, when one changed course and forced the three next to it to swing away in order to avoid a domino-like series of collisions.

Speed can also be deduced from trackways. Whether they moved singly or in groups, or whether they were young or mature, the *Amblydactylus* dinosaurs moved at an average speed of somewhat less than 5 kilometres per hour. Solitary animals showed no evidence of nervous activity. They did not mind getting wet, and some animals even entered salty water. One trackway was left behind by an animal sculling along on the tips of its toes in water about 2 metres deep. It would push with one foot, float for several metres to one side, then push with the other foot and float for an equivalent distance to the opposite side. Trackways made in deep water show little indication of herd movement. Of course there are no tracks from immature animals, for they were unable to reach the bottom with their feet.

Some of the narrow-toed footprints of carnivorous dinosaurs were made by small, chicken-sized theropods (*Irenichnites*). Others belonged to carnivores weighing perhaps 100 kilograms, which often hunted in packs of a half-dozen adolescent-to-adult animals (*Irenesauripus mclearni*). Trackways of large theropods, standing 2 metres tall at the hips (*Irenesauripus acutus*), occur singly or in pairs, and the animals apparently preferred a more solitary mode of hunting. Carnivore prints have been found in deep water, so that herbivores were not entirely safe there. The trackways of flesh-eating theropods are too abundant relative to those of the herbivores for the two groups of trackmakers to have lived in ecologic equilibrium. The carnivores evidently preferred the open mud-flats because they patrolled a larger area in less detail than did the plant-eaters. They were also more active, moving at an average speed of $7\frac{1}{2}$ kilometres per hour. The slowest speed indicated by a carnivore trackway is equal to the fastest speed recorded for that of an *Amblydactylus*, and the fastest was rivetted into the substrate by an animal running at a speed of over 16 kilometres per hour, close to the maximum speed attainable by a human runner. There is a frightening element of single-mindedness and purpose expressed in the swift, straight trackways of carnivorous dinosaurs.

Trackway of a carnivorous dinosaur:
Gething Formation, Peace River,
British Columbia

The oldest known bird tracks (*Aquatilavipes*) occur in busy profusion on a lithified mudbank. They were made by birds of about the size of a killdeer, but in view of their antiquity these birds may have possessed teeth. The pattern of tracks authentically recalls the social hyperactivity we have come to associate with small feathered folk. Trackways of other animals are not common. There are a few belonging to a small iguanodont (*Gypsichnites*). A single trackway of a large quadruped (*Tetrapodosaurus*), waddling along at just under 4 kilometres per hour, was probably made by a heavily armoured nodosaur. One turtle left its mark in the record – its velocity has not been calculated, but there is no question that, of all the reptiles there, its kind ultimately won the race with time. Many of us have exchanged glances with a turtle, but who has done the same with a dinosaur? Even more than in England, bipedal dinosaurs dominated the trackmakers almost to the exclusion of four-legged forms.

The trackmakers lived in a wetlands, where dense swamp forests and shallow wadable lakes abounded. But the Peace River area was probably close to a zone of change in continental vegetation. Unlike in the far north, cycad-like plants were both abundant and represented by a number of different varieties. Bushy, ancestral flowering plants were expanding gradually into the northern lands and by this time had already reached less swampy environments in southern Canada. Further to the south the lands were warmer and generally drier. The animals were different too.

An Exception to the Rule

The beginning of Cretaceous time (between 144 and 119 million years ago) has been described as a gap in the history of dinosaurs on this continent. This is not quite true. Near the Black Hills of South Dakota, between about 130 and 125 million years ago, there was a lowland in which stream, swamp, and lake sediments were trapped (lower Lakota Formation, 70 metres in thickness). Cycad-like plants (bennettitaleans) grew there in variety and profusion. Some many-branched forms resembled the modern Joshua trees of the southwestern United States, but others had a single, barrel-shaped stem and looked like giant pineapples. These odd plants did not survive the age of reptiles. Although they were a long-lived group, they appear to have flourished particularly in semi-arid lands during early Cretaceous time. Their odd trunks and leathery but fern-shaped leaves must have lent an exotic and haunting appearance to early Cretaceous landscapes, particularly in the oblique light of the setting Sun. Some bore flower-like cones on their trunks, while in others the cones never opened but turned into fruitlike pods with the seeds inside. Insect tunnels have been found in their petrified trunks, and a few of the squatter forms collapsed as they decayed so that the crowns were found fossilized within the stems. There were the ubiquitous fields of ferns, and cedar-like conifers may have tolerated soils moistened by alkaline lake water.

A few dinosaur bones have also been found. A small herbivorous running dinosaur related to *Hypsilophodon* is represented by a thigh bone. Although the animal was four times heavier than its English counterpart, it still weighed only about 60 kilograms. A skull of a larger bipedal plant-eating dinosaur was recently discovered that belonged to a previously unknown dinosaur. This animal probably weighed somewhat less than 1000 kilograms. In another specimen only the bones of the hip region are preserved, but these may belong to the same kind of herbivorous dinosaur that produced the skull. An additional partial skeleton, representing an animal also weighing about 1000 kilograms, belonged to a squat armoured dinosaur (*Hoplitosaurus*). Although they were closely related to Wealden dinosaurs, all three of the South Dakota forms were distinct. None of them was nearly as ponderous as the Wealden sauropods or iguanodonts. Living deep within the interior of a continent, rather than near embayments of the sea, the South Dakota dinosaurs must have been adapted to a leaner environment.

Iron Swamps

The oldest continent-wide Cretaceous dinosaur fauna known in North America belongs to an epoch lasting between about 113 and 105 million years ago, when dry land still intermittently linked the Pacific and Atlantic coasts. The dinosaur faunas seem not to have been extraordinarily rich, and the animals tended to be small, by dinosaurian standards. The faunas have a peculiar character all their own, which perhaps can be typified by the presence of the small sauropod *Pleurocoelus*. But when one thinks of the time of *Pleurocoelus*, images of ornithopods that looked like lizard-cows, armoured nodosaurs, and utterly rapacious small carnivores immediately come to mind. So also do those of some of the most fascinating dinosaur trackways known. Sediments of this age have the potential to reveal many new and interesting dinosaurs.

Dinosaurs belonging to the age of *Pleurocoelus* were first collected in Maryland, northeast of Washington, DC, from iron-rich deposits (Arundel Clay, 30 metres thick) that were mined from colonial times to just before the First World War. The metal occurs as iron carbonate in hard nodules and ledges. These in turn lie within dark clays rich in plant debris and associated with low-grade coals. Stumps, branches, and leaves were buried in fetid muck accumulating in old swamp-filled stream valleys and tidal marshes by the sea. Bordering the gloomy forests of ferns and swamp conifers, and only where the passage of streams allowed abundant light to reach the ground, were scattered bushes bearing small, flat leaves instead of needles or fronds. These were primitive flowering plants, which with the passage of Cretaceous time would rise to dominate world floras. They were organized in a highly efficient manner, so that water was rapidly transported through the stem to all plant tissues, and if growing buds were eaten new ones would quickly become active to ensure the plant's survival. They possessed a unique method of pollination, ensuring that food would be stored in seeds only if the seeds had been fertilized. Perhaps a dozen different kinds survived along stream banks, and they were very exposed to destruction by erosion or overgrazing. At the time they probably resembled peculiar many-branched cycad-like plants, and their great vigour and evolutionary potential may not have been immediately obvious.

The most common animal seems to have been *Pleurocoelus*. The relative abundance of its scattered remains recalls the late Jurassic Morrison fauna which was dominated by so many different kinds of giant sauropods. But only one sauropod is known from the swamps of Maryland, and it was not large. More than half of the *Pleurocoelus* bones came from young animals weighing about 500 kilograms and less than 5 metres long. Isolated skeletal fragments indicate the presence of animals weighing 4 metric tonnes and attaining a length of 10 metres. *Pleurocoelus* was a slender, long-legged animal, and all four limbs were of about equal length. The bones at the base of the hand were particularly long, and increased the height at which the chest was carried above the ground. Its tail was short and light, and its neck was only moderately long. *Pleurocoelus* differed from the Morrison giraffe dinosaurs in somewhat the same manner as the shape of the forest-dwelling okapi differs from that of its giraffe cousin.

Other dinosaurs are represented only by isolated teeth and bones. A few teeth of a nodosaur (*Priconodon*) and one of an ornithopod (*Tenontosaurus*) have been identified. Among the remains of carnivorous dinosaurs are bones of a tiny raptor, a primitive ostrich dinosaur (*Archaeornithomimus*), and a theropod weighing on the order of 1.5 metric

Cloverly Formation, Crooked Creek, north of Lovell, Wyoming.
Morrison Formation: light sandstones at base of exposures; Cloverly
Formation: dark grey, light grey, and maroon shales in face of escarpment
(*Tenontosaurus* skeleton discovered near centre); Sykes Mountain
Formation: rusty-coloured beach sandstones on ridge near top; marine
Thermopolis Shale: well-bedded grey strata overlying Cloverly

tonnes. Similarly, a few elements indicate the presence of fish, turtles, and crocodiles. Our best hope for increasing our knowledge of the dinosaurs that once lived in the Arundel swamps now seems to lie in maintaining a vigilant watch on excavations for large buildings between Baltimore and Washington.

Dinosaur Steppes

Volcanoes erupted in central Idaho and Arizona during the age of *Pleurocoelus*. Between the centres of volcanism huge compressional forces generated by the gradual collision of the Pacific Ocean floor with the North American continent caused great thicknesses of stratified rock to rupture and slide over each other. Lakes formed on the plains to the east of the ruptured terrain, from eastern Idaho and wes-

tern Wyoming into central Utah and Nevada. The lakes were populated with freshwater fishes and invertebrates, and pollen from surrounding vegetation blew into them. The skeletons of fish-eating flying reptiles have been collected in sediments deposited in similar lakes in Argentina and China, but none has yet been found here.

The seaway lying between the low western mountains and the shield of ancient rocks in northern Canada continued to expand to the south into southern Alberta and Saskatchewan. Most of the streams draining the western interior basin of the United States flowed north to empty into this gulf, passing around island mountains rising above the plains in central Colorado. As the lands along the perimeter of the vast basin were uplifted, Morrison sediments were attacked by erosion and redeposited on the northern interior plains of Wyoming and Montana (Cloverly Formation, 200 metres thick). At first, gravels from the compressed terrain to the

115 million years ago

In the western US states, a well-developed system of rivers spreads eastward, from rugged, hilly terrain, across semi-arid interior plains.

west flooded the northern plains. As these terrains were eroded, stream sands were swept in from the northeast and east, although ash falls from the western volcanoes were carried onto the plains as well. Soda deposits in the soils indicate that the land dried out completely during the dry season.

Although a relatively large number of dinosaurian skeletal fragments have been collected, most of them can be identified as one of three varieties (*Tenontosaurus* and *Sauropelta* were herbivores; *Deinonychus* was a carnivore). Evidently not many different kinds of dinosaurs were able to prosper on the arid plain, simply because the environment must have been too hostile. The two major herbivores were adapted to browsing in the crowns of low bushes. A windy plain can be visualized, covered with a sparse mantle of cedar-like plants. Everything carries a veneer of fine dust from the red or grey earth. Broad soda pans produce shimmering mirages, the benign illusion of which is shattered by the vertical columns of dust devils wandering across the surface of the 'lake.' The afternoon sun is so hot that it is difficult to comprehend how anything could bear to move.

Yet what looks like a long-legged lizard of bovine proportions is moving in a deliberate manner across a dry wadi. The *Tenontosaurus* is walking on its forelimbs as well as its more muscular hind limbs, carrying a long, bladelike tail well off the ground behind. The parched surroundings suggest that the tail is much more useful for capturing warmth from sunlight when the sun is close to the horizon and the air is cool than it ever would be for swimming. The pebbly skin is an orange and blue-green patchwork that renders the animal nearly invisible in the bushlands. Its rather peculiar head combines attributes of a goat and an iguana. In each jaw the teeth are locked together loosely in a chewing battery resembling that of *Iguanodon*. However, the absence of a thumb spine, the weakly constructed wrist, and the presence of four toes in the foot imply a closer relationship to hypsilophodonts. The animal is over 6 metres long and weighs 900 kilograms and is thus much larger and less fleet than the English *Hypsilophodon*. *Tenontosaurus* tends to be solitary, for its remains are never found in bonebeds that would suggest herding. Yet on the flatlands they outnumber those of all other kinds of dinosaurs combined. The small dinosaur enters a group of bushes that were tipped downstream by currents during previous times of high water. A turtle dives into a stagnant pond on the opposite edge of the clump.

In another part of the flatlands the water table is higher, and the moist soil supports a denser stand of small bushes, interspersed with ferns and small trees. A group of four barrel-bodied dinosaurs is stripping fronds from large ferns, and the shaking plants accurately mark the passage of otherwise invisible browsers. They are shorter than *Tenontosaurus*, being only a little over 5 metres long. Though over half of their length is taken up by a long, almost ratlike tail, the animals are very heavy and weigh some 2500 kilograms. That much protein concentrated in one bulky herbivore identifies it as a very attractive food item and underscores a defence requirement. The animal, called *Sauropelta*, is accordingly heavily armoured. It is the most completely known nodosaur in North America. There are round, bony discs imbedded in the skin over the hips, and the tail is covered with several rows of small conical spines. The most spectacular armour is concentrated along the sides of the animal in a row of large, sharp spikes extending from the shoulders to the head. The skull is also heavily constructed. It is apparent from the distribution of armour on *Sauropelta* that carnivorous dinosaurs were selective in their target areas. They would have preferred to seize their prey by their necks, had not the spines there been so large.

Nodosaurs were often among the more common dinosaurs in faunas during, and immediately before and after, *Pleurocoelus* time. The oldest nodosaurs occur in middle Jurassic faunas of England, and they are well represented in the Wealden. One left behind a trackway in the Peace River gorge in British Columbia, and another left a skeleton subsequently named *Hoplitosaurus* near the Black Hills of South Dakota. It was smaller than *Sauropelta*, but even more stocky. The *Priconodon* teeth in the *Pleurocoelus* swamp of Maryland indicate that this nodosaur was also distinctive. During early Cretaceous time nodosaurs seem to have been the only armoured dinosaurs in North America, and they lived there in variety.

The setting Sun enhances the red soils and the red dust on the leathery fronds of cycad-like plants. A young *Tenontosaurus* nestles its body into warm sand, partly sheltered by roots dangling down from the eroding bank overhead. A small carnivore darts across the stream bed, and the force of the collision of the bodies of the two animals pins the young herbivore, struggling to rise, against the stream bank. The carnivore slides the claws of its hands into the back of the tenontosaur and seeks to immobilize the flailing head in its jaws. Using the weight of its body to pull the herbivore to one side, it rips out the viscera with alternating rakes of the curving talon on the inside of each foot. The writhing slowly abates, as viscera are coated with dirt and bright red blood coalesces into sandy balls. The sunlight vanishes, and the small carnivore (*Deinonychus*) releases its prey and begins to devour the interior of the body cavity. Radiant energy from a star, the Sun, having been captured by green plants whose foliage was consumed by a herbivorous dinosaur, flows into the body of a carnivore in an ecological process that can be described with numbers by a myopic paleontologist many tens of millions of years later. This act of energy flow, endlessly consummated in nature, epitomizes the marvel and horror of the living world. Its antithesis is the microscopic marvel of conception. *Deinonychus* played its role well, but without a thirst to comprehend. Its bones indicate an animal 60 kilograms in weight and 2.5 metres in length. Broken teeth are often found buried with juvenile *Tenontosaurus* skeletons. The reason for the association is evident.

A few specimens of *Pleurocoelus* continue to browse in the twilight among the branches of a thicket of small trees, growing from a river bed that holds running water throughout the year. One large animal weighing 13 metric tonnes strikes a piece of partly buried driftwood. The limb flops out on the surface of the sand and simultaneously ejects a soggy tail vertebra of a large theropod. The bone falls to the side, one half bleached by the sun and the other blackened by wet sand. Several ostrich dinosaurs nervously trot to a more open space near the centre of the river course, while an immature long-legged egg-stealer (*Microvenator*) cocks its head to inspect thoughtfully a mound of sand. In the gathering dusk the form of a small, fleet hypsilophondont (*Zephyrosaurus*) is barely visible as it makes its way toward a small pond. The welcome darkness of a cool night spreads over a strange land, enforcing rest.

To the north in Montana the flatlands became wetter, and plant cover grew more densely. The sagebrush-like conifers and cycad-like plants give way to tree ferns, taller conifers, and ginkgos. As the margin of the northern sea was approached the soils were frequently covered with water

Details of *Tenontosaurus* (left)
and *Sauropelta* (right), Wyoming,
100–110 million years ago,
during early Cretaceous time

up to 1 metre deep. The swamp forest was itself stratified, with a canopy of conifers on top, a second storey of tree ferns, and, where the soil was exposed, a ground cover of ferns and bryophytes. Some of the swamp trees shed their needles during the dry season, as bald cypresses do today. The sedimentary sheet is thicker here than further south (Kootenai Formation, 379 metres thick) and shows evidence of lakes and coal swamps. Not many bones of dinosaurs have been seen, but those that have been reported belong to sauropods. A pair of small jaws imbedded in red siltstones (*Toxolophosaurus*) evidently belonged to a sphenodontid, a group of lizard-like reptiles that survives today in New Zealand.

The southern flatlands had initially been dry, and few dinosaurs other than *Tenontosaurus* inhabited them. The armour plates and bones of *Sauropelta* groups overcome by flash floods were occasionally washed out into the arid central flatlands, as were water-worn logs and broken branches. The dry climate with its seasonal rains was preferred by lungfish and by tiny, insect-eating primitive mammals (triconodonts). However, the surface of the land slowly subsided toward sea level, and tidal flats began to expand across Montana. Lakes and marshes spread south in front of the Canadian sea into Wyoming. With the approaching sea came heavier rains, which penetrated more deeply into the interior. Plant life flourished, and coal swamps formed northwest of the Black Hills in Wyoming. Parasol-like ferns (*Weichselia*), formerly common in the Wealden of Europe, spread across the region, and from time to time the broad, compound leaves of primitive flowering plants fell into stream-side ponds. Dinosaurs became more abundant on the greening savannahs, and populations of *Pleurocoelus* and *Sauropelta* particularly prospered.

A Bone on a Delta

Far to the south, in what is now the southeastern corner of Arizona, a tremendous series of deltaic sediments was accumulating under a broad western embayment of the Gulf of Mexico (Shellenberger Canyon Formation, 4500 metres thick). The lower third of the sedimentary sequence was deposited on a delta floodplain that later passed beneath the surface of the sea. These sediments have yielded abundant fossil tree trunks, washed down from rugged terrain to the north, and a single dinosaur thigh-bone, tentatively identified as belonging to *Tenontosaurus*. However, it was derived from an animal weighing over 1500 kilograms and thus much heavier than *Tenontosaurus* specimens from the northern flatlands. Was the animal different?

Trackways in Lagoons

The arid plains formed a broad isthmus separating the Canadian sea from the tropical waters of the Gulf of Mexico. Southern Texas lay beneath shallow, blue-green lagoons, which extended from a low shoreline west of the present highway between Dallas and San Antonio for hundreds of kilometres to the southeast. A huge barrier reef separated the seaward margin of the lagoons from the deep blue waters of the gulf. Sweeping around large offshore banks in Chihuahua and Coahuila, the reef paralleled the southwestern coast of the United States for thousands of kilometres. The waters of the lagoons were clear, for streams carrying sediments from the interior of the continent flowed to the north. They were somewhat saltier than the open ocean because the climate was hot and evaporation rates were high. Organic mats of carbonate-secreting organisms were widespread beneath the shallow, brightly illuminated waters, and lime slowly accumulated on the floor of the lagoons. After heavy rain squalls, varicoloured clays were washed into the lagoons, carrying with them pieces of wood and bones. Land-derived sands and clays in the west thus merge with shallow-water limestones in the east (Antlers Sands–Glen Rose Limestone, 120 metres thick).

Washed-in plant material was often buried in pure limestone. Although conifer twigs and cones predominate, wood can also be very abundant, and trunks measuring up to 60 centimetres in diameter are preserved. The conifers are peculiar in that the cuticles of their leaves are very thick. In some, the pores for gas exchange are deeply sunken. These features are typical of plants adapted to living in hot, dry climates, or on salty soils. Most of the pollen grains preserved in the sediments belong to thick-cuticled conifers, and their great abundance suggests that the parent trees grew in stands dominated by single species at any particular site. These trees may have inhabited salt marshes at the edge of the lagoons, as mangroves do today. They are members of a large group of extinct conifers, the cheirolepidiaceans, various members of which grew in a wide variety of tropical environments.

Cycad-like plants were less extensively preserved, perhaps because their fronds decayed while remaining attached to the stems. Some 7 different varieties are known, which as a group so closely resembled plants then growing near the Blacks Hills in the interior that they might have been in the same forest. Preserved stems of branched forms measure up to 11 centimetres in diameter, with the bases of dropping fronds still attached, while those of globular forms attain diameters of 50 centimetres. Three stems belonging to the latter variety had grown together in a group and were fossilized still attached to one other. The crowns of these plants bore as many 30 fronds, each containing some 50 leathery leaflets. Both branched and globular forms may have preferred better-drained, sandy soils. Fern fronds and spores are quite uncommon, and the plants must not have figured importantly in the vegetation of the region. Perhaps the climate was too hot to permit them to flourish as they did in cooler, more temperate climates to the north. The remains of primitive flowering plants are only slightly more common than those of ferns.

A 1-metre bed of white estuarine sand in north central Texas has produced a few large and small carnivorous dinosaur teeth, as well as foot bones and many bits of ossified tendon of *Tenontosaurus*-sized herbivores. This in itself is not remarkable, but the bed also contains countless tiny bones belonging to an array of small animals from fresh, brackish, and salt-water environments. There are scales, vertebrae, and teeth of sharklike fishes, deep-bodied, shellfish-eating fishes with enamelled scales, and primitive bony fishes. There are bones from many different varieties of salamanders, frogs, lizards, turtles, and small crocodiles. All these were discovered as a by-product of intensive sifting of tons of sand for the diminutive teeth of primitive mammals. These were recovered in much smaller numbers, and most belonged to long-extinct, archaic groups. Teeth of seed-eating and insect-eating forms occur in about equal numbers, and among the latter are a very few that can be included within the remote ancestry of nearly all living mammals.

Microvertebrate remains have been collected from many of the near-shore sands of north-central Texas, and their abundance contrasts greatly with their virtual absence in the sediments of the arid northern interior. Lungfish teeth, which are relatively common to the north, are among the rarest of fish fossils along the gulf coast. The presence of great concentrations of small bone argues for the existence of rich lowland vegetation behind the lagoons.

Most of the southern dinosaurian skeletal material was collected from sites in north-central Texas and southeastern Oklahoma. It occurs usually in sandy strata deposited near

Pleurocoelus

EARLY CRETACEOUS, ABOUT 110 MILLION YEARS AGO

Pleurocoelus specimens walk along end of partly exposed dead reef,
on Texas Gulf coast; another wades across shallow lagoon.
Primitive flowering bushes grow beneath conifers.

the landward side of lagoons, although one *Pleurocoelus* specimen was buried in limestone lying on the flank of a reef. The dinosaurs were similar to those of the northern flatlands, but they differed in relative abundances. There were no nodosaurs, and their absence is as striking as is the paucity of ferns in the regional vegetation. One is tempted to wonder if a special ecological relationship existed between nodosaurs and ferns. *Pleurocoelus* was relatively more common than in the northern interior, but less so than in the eastern swamps. No immature specimens are known from skeletal material, but trackways demonstrate the presence of a broad range of body sizes. Some footprints were made by gigantic animals weighing more than 40 metric tonnes.

A number of good *Tenontosaurus* skeletons have been unearthed, but, as on the northern flatlands, the animals were relatively rare in moister environments. The population living near the Gulf Coast may have belonged to a different species, for the skulls appear to be lower and possess larger nares. If the differences prove to be real, perhaps the inflated narial region served as a heat exchange to cool the brain, as in mammals inhabiting hot environments today. Other plant-eating dinosaurs were present in the fauna, but are incompletely known. Tiny and very simple teeth suggest the presence of primitive fabrosaurids. Some large limb bones could belong to iguanodonts but also resemble those of hadrosaurs (duck-billed dinosaurs) in their slenderness. A few large bones were derived from at least one additional, as yet unidentified herbivore.

There were also small raptorial carnivores, perhaps related to *Deinonychus*, and a single, isolated foot that may represent an animal allied to the ostrich dinosaurs. The most spectacular carnivore remains so far collected belong to a 9-to-10-metre-long carnivorous dinosaur that weighed about 1.5 metric tonnes. *Acrocanthosaurus* was a rather typical theropod in most respects, but spines nearly half a metre long projected above the backbone in a single row extending from the neck to the base of the tail. Some of the spines are so slender that it is difficult to imagine them supporting anything but a web of skin. Their trackways indicate that theropods were active creatures, and in warm environments a heat-radiating web may have been useful in allowing an animal to dissipate heat after running, thereby avoiding death through heat prostration. Carnivorous dinosaurs with similar webs on their backs are known from southern England and Egypt.

The shallow lagoons of central Texas were a nearly ideal setting in which to record the footprints of dinosaurs. The vigorous growth of offshore reefs isolated the lagoons from the open sea, and slight changes in sea level could cause huge changes in the distribution of limey mud-flats and back reef water. An approaching hurricane could drain the lagoons. Dinosaurs wandered across the flats when the water had temporarily drained away, the mud squeezing up into low ridges around their feet as they strode. Some tracks were filled quickly with mud washed from the land as the approaching storms made landfall. Others tracks remained, and were filled slowly over many weeks. Fragments of these once impressive vistas can be seen in Dinosaur Valley State Park, located in a bow of the Paluxy River 5 kilometres west of Glen Rose, Texas. When the river is low, and running clear but slightly milky, the giant forms of dinosaur footprints can be seen imbedded in limestone just below the surface of the water. Few sights can ever bring the dinosaurian world more stunningly to life.

The trackways preserved in Texas limestone belong to only three major varieties, but the tableau they paint conveys a power unequalled elsewhere on the globe. Some tracks are of the same form as those in the Peace River Canyon of British Columbia, and the same names have been applied to them. Broad three-toed tracks measuring 30 to 40 centimetres in length (*Gypsichnites*) were made by unknown herbivores resembling iguanodonts or hadrosaurs, but not by the four-toed *Tenontosaurus*, well known from skeletal fragments. The tracks of slender, three-toed carnivores are often tipped with claw impressions (*Irenesauripus*) and are logically attributed to *Acrocanthosaurus*. The sauropod tracks are unnamed but can be identified as those of *Pleurocoelus* through the fortunate discovery of the hind limb of an animal that had died while its leg was mired in a vertical position. Unlike most sauropod skeletons, the foot bore four claws, as do the animals that made the Texas sauropod tracks. These trackways show further that *Pleurocoelus* lacked claws altogether on its forefeet, while most sauropods retained a thumb claw. Many of the famous trackway sites in Somervell, Commanche, and Bandera counties in Texas are no longer available for viewing, although they have been well described. Together with less spectacular sites, they occur in stream valleys over a broad area between Fort Worth and San Antonio.

One trackway was made by a giant *Pleurocoelus* that may have weighed well over 40 metric tonnes. The animal was moving in a regular cadence at a normal walking speed of about 4 kilometres per hour. A large *Acrocanthosaurus*, whose hind limbs were over 2.5 metres long, walked along the same trackway, and in every case where footprints from the two animals coincide, the one from the carnivore is superimposed over the print of the sauropod. Near the centre of the excavated portion of the trackway, the carnivore's

Dinosaur tracks:
contours are visible in mud beneath
Paluxy River, in Glen Rose Formation,
Dinosaur Valley State Park,
near Glen Rose, Texas.

Tracks of sauropod (TOP),
and carnivore (*Irenisauripus*) (BOTTOM),
Glen Rose Limestone,
Dinosaur Valley

cadence seems irregular, and both animals appear to have veered to the left. Had it seized the moving herbivore, and was it pulled off balance by the powerful sauropod? It was moving twice as fast as the sauropod, but this was nevertheless a normal walking speed for a carnivore.

How much time separated the passage of the two animals? Was the carnivore simply following a fairly old sauropod trackway on a routine patrol for prey? Or were both animals nearing the point of exhaustion after a long chase through extensive mud-flats? The sauropod may not have been a solitary animal. At least three other *Pleurocoelus* trackways lay to the left on a parallel course. One was of a half-grown animal, but the others were as large as the first. Additional trackways lie beneath the centre of the Paluxy River. The four visible sauropod trackways are spread over a 20-metre width so that if the animals were travelling together it was a loosely knit group. Was the carnivore then following a herd rather than an individual? The stratum on which the trackways are preserved passes into a hillside, where they were too deeply buried to be excavated by their discoverers. Human curiosity will probably ensure that the spoor of these dinosaurs will one day be pursued into the hill.

At another locality more than thirty bipedal carnivores crossed a mud-flat in a variety of directions. The mud-flat was exposed to the air, for the imprints were made over drying cracks in the mud. Most of the tracks were only about three-quarters the length of those of the *Acrocanthosaurus* that was following the sauropod trail, and the animals were moving at speeds of between 4 and 11 kilometres per hour.

Elsewhere, a series of prints from the forefeet of a giant *Pleurocoelus*, probably weighing in excess of 40 metric tonnes, was impressed into the soft limestone. Only one hind foot made contact with the ground, leaving its print at a point where the animal altered its course. The right hind foot was used to push the animal to the left. Its hindquarters were thus floating, but they could have floated only if the animal were wading in more than 2 metres of water. The fact that the animal was sculling along on its forelimbs implies that they were at least as long as the hind limbs, in conformity with what is known of the skeletal anatomy of *Pleurocoelus*. The animal was not in a hurry, nor was it in an area where abundant vegetation was likely to be nearby. Perhaps it was simply cooling off (or thermoregulating) in the way elephant and hippo do during the heat of an African day.

A classic locality, which is as worthy of becoming a national monument as the quarry face of Dinosaur National Monument in Utah, occurs in Bandera County, west of

San Antonio. It is virtually a script for a moving picture of dinosaurian behaviour, and a sequence of events can be reconstructed by opening the mind's eye while simultaneously examining the spacing, orientation, and superposition of the footprints. A large carnivore sculls through 2 metres of water in a shallow lagoon on the tips of its toes, moving south. Later, the lagoon drains to the east, possibly because a distant but approaching storm causes water levels to fall. A small carnivore sprints to the east at a velocity of nearly 20 kilometres per hour. Did a stranded fish catch its eye, and was it trying to outrace competitors? Two large carnivores walk slowly, at a speed of about 6 kilometres per hour, toward the receding water. Their trackways are about 10 metres apart.

Then, perhaps sensing a change in the weather, a herd of 23 head of *Pleurocoelus* appears from the east, walking toward the western shore. None of the animals weighs much more than 20 metric tonnes, and the smallest individuals probably weigh about 2.5 metric tonnes. All of them, large and small, are walking at a speed of slightly less than 4 kilometres per hour, and all of them are holding their tails well off the ground. We could easily walk as fast as they do. The herd is divided into a southern group, composed of 7 individuals, and a northern group, of 16 individuals. The centres of the groups are separated by about 9 metres, and all animals are bearing in the same direction. The northern group is led by a very small animal; a slightly larger juvenile is walking near the centre of the southern group. Four animals in the northern group are walking along in single file, stepping in each other's footprints. The herd passes us in about one minute.

After the passage of the sauropod herd, a large carnivore crosses the trackways, bearing south at 7 kilometres per hour. The air becomes charged with electricity and gusty as the storm approaches. Two more large carnivores walk rapidly to the east at over 8 kilometres per hour. The storm arrives, and rising tides swell into the lagoon. Driving rains pour on the land, and sediment begins to settle out of the runoff entering the lagoon. The trackways begin to fill with mud . . .

Theropod tracks are being studied at other localities. More than 20 percent of the carnivores at one site were running as they made their trackways. One currently holds the record for the world's's fastest dinosaur. It was moving at 43 kilometres per hour, faster than the fastest human racer, but still well below the speed of a racehorse. The trackways in the lagoons provide food for thought. Those of carnivores are solitary and oriented in many directions, suggesting that these animals were wide ranging and did

not hunt in packs. They depended on speed for defence, for the trackways of smaller animals often indicate higher speeds than those of the larger ones. The sauropods preferred to travel in groups, and the slowly moving herds must have provided some protection (there is no evidence that the young animals were located in the centre of moving herds). None of the carnivore tracks was following or bearing in the same direction as the Bandera County herd. The *Pleurocoelus* specimens from the Maryland swamp seem, on average, to have been smaller than the smallest individuals in this herd. Perhaps young animals sought the shelter of swamps.

Heavy Traffic

In the southwestern corner of what would one day become Arkansas, salts crystallize out in brine pools on sun-baked mud-flats. The mud-flats form a tidal corridor a few kilometres wide between ancient eroded mountains to the north and shallow lagoons to the south. Countless footprints from *Pleurocoelus* herds blanket the surface of the corridor. The prints are often much too dense for individual trackways to be identified. A typical animal passing by weighs on the order of 20 metric tonnes and walks at the relatively high speed of 4.7 kilometres per hour. These are sauropods on the move, and there are no young animals in the herds. Indeed there is no visible animal life on the acrid, torrid corridor, other than mature sauropods. The trackways lead both to the east and to the west. A few even cross the others at right angles. Why are so many sauropods walking resolutely in opposite directions? Are they on a daily journey between water-holes and feeding grounds?

The Flooding of the Interior

Between 105 and 98 million years ago, a gradual increase in the speed of collision of the Pacific floor with the North American continent induced renewed oversliding in old sedimentary strata in eastern Idaho. During the same interval the Canadian seaway advanced far to the south. The northern interior basin rapidly fell beneath sea level, and the surface of the old arid flatlands was scoured by waves. Deeper shelf waters drowned the lagoons of Texas and Arkansas, and the Gulf Coast advanced through Oklahoma and Kansas. A strait opened near the eroded roots of hills that had formerly projected above the flatlands in central Colorado. An expanding oceanic basin in the Arctic was thus linked to another expanding oceanic basin in the Gulf

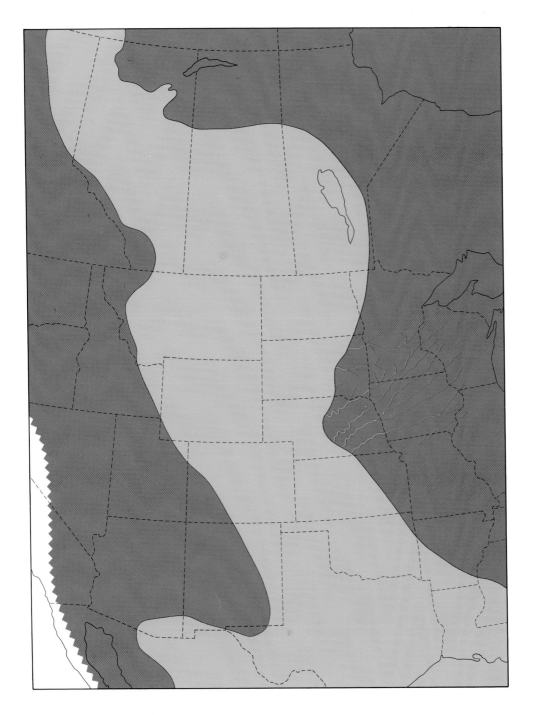

110 million years ago

Shallow arctic seas and Gulf of Mexico meet in Colorado. River valleys southwest of Great Lakes are eroding into now-ancient rocks.

Near Dalton Well Quarry,
eastern Utah.
Cedar Mountain Formation,
exposed where fin-backed
iguanodonts were discovered

of Mexico through a shallow midcontinental seaway. The seaway was 4000 kilometres long, and 1000 kilometres wide along the present border between Canada and the United States.

Silts and shales deposited under the deepening sea floor in Wyoming (Thermopolis Shale, 30 metres thick) contain the bones of short-necked plesiosaurs with long heads. Occurring in the same strata with the skeletons of these fish-eating marine reptiles are occasional bones of crocodiles. The discovery of two skeletons of armoured dinosaurs buried in these strata is an indication that the coastline was not far away. The bloated carcasses were evidently washed out to the sea, where they sank. *Nodosaurus* was a relatively large animal, measuring over 5 metres in length and weighing about 1.7 metric tonnes. It possessed longer forelimbs and shins than other nodosaurs. The coincidence that *Nodosaurus* is the only dinosaur so far identified in these marine sediments further suggests that nodosaurs preferred moist lowlands near the sea.

Alluvial fans spread from rugged hills in central Idaho, merging to the east with a narrow floodplain bordering the shallow sea (Thomas Fork and Wayan formations, 600 and 1000 metres thick, respectively). Soils between streams were well drained, and the water table was low most of the year.

Ornithopod dinosaurs, including perhaps *Tenontosaurus*, came to lay their eggs in the well-aerated, sandy red soil. Eggshell fragments are abundantly preserved, and the curvature of the pieces indicates that some eggs were as large as 8 centimetres in diameter. Many-crowned *Tempskya* ferns grew beside streams inhabited by snails, fishes, and soft-shelled turtles. A few isolated teeth of *Tenontosaurus* and nodosaurs show that these plant-eating dinosaurs frequented the more heavily vegetated stream banks as well. Eggshells have not been found in the sediments on the arid flatlands, and the dinosaurs may have preferred to lay their eggs in higher regions, where the young could feed on freshly grown foliage nourished by perennial streams.

The coastal floodplain (Cedar Mountain Formation, 140 metres thick) widened to the south across Utah. Its silts and clays are tinted with delicate hues of green, purple, and red, interrupted occasionally by yellow lenses of river sand. In carbonaceous layers representing the site of former swamps, *Tempskya* trunks can be found buried upright in their position of growth. Conifers grew in well-watered areas, accompanied by flowering plants that by now had also attained the proportions of trees. The variety of pollen grains preserved in the sediments indicates that flowering plants were becoming adapted to a wider range of habitats.

Sandstone blocks
(Cedar Mountain Formation),
near Dalton Well Quarry

The bones of the animals that once lived in this region are only in the initial stages of being collected and studied. Skeletal scree concentrated on the bottom of rapidly flowing channels includes bones of fishes and crocodiles, as well as teeth of mammals and large dinosaurs. Teeth from embryonic dinosaurs and eggshell fragments suggest that streams cut into fresh nesting sites and the eggs were spilled into the current. *Tenontosaurus* frequented the plain, as did *Iguanodon*. A thigh-bone with a very straight shaft may document the presence of hadrosaurs. Hadrosaurs, or duck-billed dinosaurs, were descended from iguanodonts, but their legs were relatively longer, and their dental batteries were more compact and had more teeth. They appear to have been better adapted to walking long distances for food and evidently more thoroughly chewed the plant materials they cropped. The skeletons of hadrosaurs give the impression of belonging to more active animals than do those of iguanodonts. Pieces of the skeleton of a large *Sauropelta* have been collected, so that quadrupedal armoured nodosaurs

also occurred on the southern plain. The skeletal material available so far suggests that this is the oldest known skeletal assemblage of dinosaurs in North America dominated by bipeds. However, trackway assemblages dominated by ornithischian bipeds occur in strata deposited some 15 million years previously in British Columbia bayous.

Just before the end of what has formally been defined as early Cretaceous time the long north-south seaway was broken in central Colorado. The Canadian arm retreated slightly to the north and later re-expanded but did not contact the Gulf of Mexico. Sediments (Mowry Formation, 50 metres thick) deposited on the floor of this great inland bay, which completely covered the site of the old arid flatlands, are filled with volcanic ash blown from western volcanoes. Perhaps ash falls poisoned the surface waters, and the mass death of millions of fish produced strata rich in scales that are used to identify the boundary between early and late Cretaceous sediments deposited in Canada. Flying reptiles dotted the skies over the Mowry sea and contributed a few

Weathered dinosaur bone,
near Dalton Well Quarry

bones to the siliceous muds on the sea floor. Ichthyosaur bones are known from the older lagoonal limestones of Texas, and skeletons of these beautifully formed creatures have also been found in the ash-rich sediments. They had been decreasing steadily in abundance and variety since the beginning of the Jurassic Period.

The Dawn of the Late Cretaceous

Dinosaurian assemblages of late Cretaceous age in North America are world famous for their spectacular array of hadrosaurs and horned dinosaurs. At the beginning of the epoch, however, the record is exceedingly meagre, and such information as is available implies that dinosaurian faunas had not changed greatly from those that had inhabited early Cretaceous lowlands east of rugged terrain in British Columbia, Idaho, and Nevada. The time under consideration covers a span of 7 million years, between 98 and 91 million years ago. The Canadian and Gulf of Mexico arms

of the interior seaway had again contacted each other and remained in connection until the end of the dinosaurian era. Yellow, near-shore sands lined both coasts through the United States (Dakota Formation, 100 metres thick), which in the west contain a record dominated by trackways, and in the east a sparse record in which skeletal remains figure more prominently.

A trackway site lies in a throughway cut (Interstate 70) west of Denver, Colorado. Sands are exposed there that were deposited on tidal flats linking a well vegetated plain with the sea. For a brief moment, now petrified in time, numerous dinosaurs walked along an ancient coastal highway, most of them moving in a southerly direction parallel to the strand line. On the landward side, the sands have produced fossil leaves, many from broad-leaved trees, and a plant microfossil assemblage dominated by the pollen of flowering plants. Very infrequently, the sands will yield a scrap of turtle bone, a crocodile tooth, or a fish vertebra.

One interesting trackway (*Walteria*) was made by a four-

Giant monitor lizard,
Jakarta Zoo, Indonesia

footed, lizard-like animal that dragged its tail in the sand. It was very robust for a lizard, and its feet were webbed. The animal may have been an aigialosaur, related to living monitor lizards. It may also have been in the ancestry of the mosasaurs, which were to become the most abundant carnivorous reptiles in late Cretaceous seas. There were also footprints from a shore bird (*Ignotornis*) with a short web between its toes, made in communal profusion within a shallow pond. It bore a short talon-like toe on the back of its foot. A trackway, possibly of an ostrich dinosaur, was made by an animal walking along briskly at a speed of nearly 11 kilometres per hour. A large carnivore took life in a more tranquil fashion, wading on its toes in deeper water slightly offshore. The most abundant trackways belong to large bipedal herbivores and were very probably made by hadrosaurs. The ends of the toeprints appear more rounded than in more ancient Canadian trackways, suggesting a rounder, more hooflike nail and a more mobile mode of existence. Prints of the forefeet show clear impressions of the individual fingers. The animals were moving, often on all four feet, at speeds varying between 4 and 5 kilometres per hour. Some trackways parallel each other, as if they were moving in groups. Similar herbivorous dinosaur tracks are preserved abundantly in other sites of the same age in eastern Colorado. The trackway of a gigantic sauropod has been reported from one locality in the state.

Hadrosaur prints are present in ripple-marked sands exposed in Clayton Lake State Park in northeastern New Mexico. Here they were also made by animals of varying degrees of maturity. One hadrosaur had slipped and caught its balance by pressing its tail against the ground. This is the only known tail print of a hadrosaur. Another unique set of prints was made by what appears to have been a web-footed carnivorous dinosaur waddling slowly along with the shaft of the lower (metatarsal) segment of the hind limb in contact with ground. It would be fascinating to know what this animal looked like. There are many prints made by other small carnivorous dinosaurs, and a few of a much larger form. Tracks of a small carnivorous dinosaur have also been recorded in deltaic strata of the same age, far to the north, in northeastern British Columbia.

Skeletons of spectacularly large plesiosaurs (*Thalassomedon*), of the long-necked or elasmosaurian variety, have been collected from mudstones deposited off the shore of the sea in eastern Colorado (Graneros Shale, 27 metres thick). These animals attained lengths of nearly 13 metres, and the head and neck of one specimen was 7 metres long. The massive animals rowed through the water, rapidly thrusting their necks toward fish and trapping them in a basket of long teeth. A smaller reptile (*Dolichorhynchops*) of the short-necked, or pliosaurian variety occurs in limey sediments in Kansas (Greenhorn Formation, 50 metres thick). Perhaps these long-skulled animals dived on their prey like beaked whales. In the same strata was found the bone of a foot (tarsometatarsus) of the most ancient flightless diving bird (hesperornithiform) known from this continent.

Plant life flourished on the eastern coast of the interior seaway, under the influence of warm, moist climates. By this time a fundamental change in plant life had taken place. The continued abundance of fern spores in sediments suggests that fern prairies remained an important part of the regional vegetation. However, the broad leaves of many different kinds of flowering trees were deposited in great abundance along stream courses. In detail the leaves prove to be quite distinct from those of modern broad-leaved trees, but they do often resemble them in size and general outline. Even at some distance, however, it might have been easy to distinguish these ancient forests from their modern counterparts in the southeastern United States. The broad-leaved trees were low in stature, and the canopy they formed was probably pierced frequently by the dark green columns of towering conifers. Broad-leaved shrubs occasionally dropped their flowers into fine-grained sediments. As reconstructed, their fragile forms evoke the loveliness that we associate with flowering trees. One was a peculiar, magnolia-like blossom (*Archaeanthus*) measuring 15 centimetres across. Others were reminiscent of apple blossoms and sycamore flowers. One tree produced seeds in clusters (*Lesqueria*) resembling shaving brushes. The woodlands must have attracted hordes of insects during blossom time, for the flowers contained nectar and their pollen was adapted for insect transport.

Leaves of broad-leaved trees are often preserved in large numbers around the few skeletal parts of dinosaurs that have come to light in the region. A skeleton of a sauropod has been observed in sandstones of this age in northern Wyoming, and a single bone possibly of a dinosaur was collected in western Iowa. From eastern Nebraska came the end of a thigh-bone of a very large hadrosaur. The best skeleton so far excavated belonged to a small nodosaur (*Silvisaurus*), discovered in eastern Kansas. Although the specimen was a mature animal, it was only 3 metres long. It was primitive in that it still bore teeth along the beak at the front of the upper jaws, and the ribs along its back had not co-ossified with the vertebrae to support the armoured carapace on the back, as the ribs have in skeletons of other nodosaurs. Although very little skeletal material is known, nodosaurs may have remained an important element in the community of herbivorous dinosaurs. Nodosaur scutes were washed into marine strata of this age in Texas.

The most southerly dinosaur occurrence in North America consists of a single bone discovered in red siltstones of this age in Honduras. The lands of Central America were bordered on the east by shallow tropical seas and linked to the rest of North America by a chain of volcanic mountains on the west.

[6]

Sauropelta

MIDDLE CRETACEOUS,
ABOUT 100 MILLION YEARS AGO

Two nodosaurs (*Sauropelta*) browse beside
a shallow sea, in the North American
interior. The vegetation links conifers to
flowering plants and is restored after
Welwitschia (a bizarre plant, with
straplike leaves, peculiar to southwestern Africa) and *Gnetum* (a small bush,
abundant only in the US southwest).

Cretaceous Seas

Before the Chalk

The Cretaceous was a time of rapid movement of the great plates that make up the crust of the Earth. Crustal materials rose beneath mid-oceanic rifts, which are the centres of spreading between plates, pushing the rifts high above the surrounding abyssal plains. The volumes of the ocean basins were diminished by enormous submarine rift ranges, so that ocean levels rose and shallow seas in turn spread across the continents. Fully a third of the present land surface of North America was inundated. At no prior time during the dinosaurian era had so much of the continent been submerged. The broad, north-south belt of shallow sea water, which split North America in half, greatly affected environmental conditions in borderlands. The area of rich deltaic plains was greatly increased. Rains enhanced the fertility of the plains, and summers were cooled and winters warmed by winds from the sea. Dinosaurs must have flourished on them, but the sediments deposited on these lowlands were largely destroyed through erosion when the sea later withdrew from the continental interior. Sediments deposited on the old sea floor are, however, widely distributed across North America, and remains of back-boned animals can be collected from them wherever the strata are exposed. Life in the seas was as different from that of modern times as was life on the dinosaur-dominated lowlands.

The fishes that inhabited the seas as the Cretaceous began 144 million years ago were generally of modest size. Among the sharks and their relatives, those adapted to scavenging on the sea floor were most common. The bony fishes were archaic in appearance. They were typically covered with an armour of enamelled scales and for the most part would have seemed rather heavily constructed. In only a few were the bodies supported by a skeleton fully formed in bone and sheathed by thin and flexible scales. These were the remote ancestors of most modern bony fishes. Although they were not common, they had begun to diversify into the basal stocks that led to salmon and spiny-rayed bony fishes, on the one hand, and mooneyes, herrings, and ladyfish, on the other.

Tidal streams emptied into the shallow lagoons of Texas, which were crossed by herds of the sauropod *Pleurocoelous* 115 million years ago. These waters were inhabited apparently by vast numbers of tiny fishes, which constituted the prey of small sharks (lamnids), primitive sawfish (sclerorhynchids), and primitive bowfins (amiids). A few lungfish teeth were from time to time washed in during periods of high runoff from the dry plains in the interior. These, together with garlike forms (semionotids) and archaic bonefish (albulids) subsisted on freshwater shellfish. The offshore reefs hosted a variety of heavily scaled bony fishes, some with short but fusiform bodies (caturids) and others with deep bodies (pycnodonts), which, together with pavement-toothed sharks (hybodonts) and skates (platyrhinids), grazed on the molluscs they found there.

To the north, shallow seas extending across Kansas and Wyoming were inhabited by free-swimming sharks and a few deep-bodied pycnodonts. Related to the ancestral mooneye-herring-ladyfish stock were large fishes that swallowed their prey whole rather than tearing pieces from them like predatory sharks. These were the ichthyodectids that later came to dominate shallow Cretaceous seas. Ocean waters remained warm across southern Canada and supported raylike sharks (*Ptychodus*), which fed on shellfish, and free-swimming marine needle-nosed 'gars' (aspidorhynchids), ichthyodectids, and ancient representatives of the herring and mooneye groups. Advanced bony fishes

Beach near exposed reef,
Udjung Kulon Rhinoceros Reserve,
Indonesia

were more abundant in the open waters of the interior sea than in the lagoons of Texas. Fish populations were sufficiently dense to sustain ichthyosaurs. Growing to lengths of 5.5 metres, *Platypterygius* left remains that have been recovered from localities scattered between Texas and the Northwest Territories of Canada, as far west as Oregon on the Pacific coast, and at other sites scattered around the world. Their jaws resemble toothed clubs and were used to stun and impale heavily armoured fishes. Two varieties of plesiosaurs, the long-necked elasmosaurids and the long-skulled polycotylids, were relatively abundant as well.

For about 20 million years life in the sea did not appear to change greatly. A stubby kind of fish (enchodontid) derived from ancestral salmon stock, which bore powerful fangs in an overly large head and calls to mind a piscine version of a Tasmanian devil, did make its appearance in North American waters during this time. Perhaps more than any other group, enchodontids typified marine fishes during the remainder of the dinosaurian era. The bones of these adapt-

able creatures have been recovered from marine limestone, chalks, shales, white sands and greensands, and sulphurous, carbon-rich mudstones from California to New Jersey and from the Beaufort Sea to the Gulf of Mexico. But in Texas the lagoons were still dominated by bottom-dwelling sharks, primitive sawfish, and archaic bony fishes with heavy scales. Plesiosaurs of both varieties, accompanied by ichthyosaurs, sculled through the old Cretaceous seas of the midcontinent.

The history of life is replete with mysteries. Most of these are so obscure that insufficient data are available even to formulate useful questions. In the case of a few, enough information exists so that a host of alternative possibilities can be imagined. These are the questions that will soon be answered. Is it true that every 26 million years or so the biosphere of our planet is briefly disturbed, and many species simultaneously become extinct? Could these extinctions, if they exist, be produced by comet showers in the inner solar system caused by a distant stellar companion of

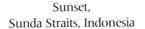

Sunset,
Sunda Straits, Indonesia

the Sun? Does this hypothetical 'death star' really exist? Assuming that waves of extinction do occur at 26-million-year intervals, previous and succeeding extinctions in the marine record predict that another extinction should also have occurred about 94 million years ago. A preliminary but global examination of available fossil evidence does suggest that an interesting number of organisms may indeed have become extinct in the seas at this time. Ichthyosaurs vanished entirely from the record, and many varieties of heavily scaled fishes declined in importance. Much will soon be learned about this 94-million-year-old 'event.'

The Coming of the Chalk

The plesiosaurs survived the extinction, as did the newly evolved bony fishes with thin, flexible scales. Within a few million years all the basic varieties of marine vertebrates typical of the interior sea during the latter part of Cretaceous time had appeared. However, the transition was character-ized by evolutionary innovation. Primitive mosasaurs, arising from lizards related closely to modern monitor lizards such as the Komodo Dragon, entered the sea. They were relatively small and evidently much less abundant than they subsequently became. The oldest known birds, which possessed powerful chest muscles and were thus capable of sustained flight, appeared. They retained teeth in their jaws (*Ichthyornis*) and were not ancestral to any modern birds. A single bone has been recovered from marine limestone that belonged to a flightless diving bird related to *Hesperornis*. Sea turtles (*Desmatochelys*) were widespread, occurring in strata as widely separated as Arizona and Minnesota. Turtles, fishes, birds, mosasaurs, and plesiosaurs all drew their nourishment from a food chain based firmly on oceanic plants.

What does an oceanic plant look like? Land plants are slowly growing organisms that can reach great ages and enormous dimensions. They can dominate landscapes to such an extent that animal life is hidden by them. Trees and grasslands assume an unchanging quality that lends stability to our lives. Oceanic plants, in contrast, are usually tiny algal cells that float with the currents and have little more visual impact than a slight alteration in the colour of the water. Their great importance lies not in spatial dimensions but in time. When light and nutrients are in appropriate supply the cells multiply at prodigious rates and provide a supply of plant tissue for food that is comparable to that produced by their much larger relatives on land. Almost paradoxically, plants that are invisible to the naked eye nourish the conspicuous animal life of the sea.

Like land plants, the single-celled floating plants of the sea have evolved through geological time. A new variety appeared in marine waters during the Jurassic. They bore tiny discs of calcium carbonate in the cell wall, called coccoliths. As the cells grew, or divided, coccoliths were shed and fell to the ocean floor. The coccolith algae gradually increased in abundance and variety. By the time of the extinctions 94 million years ago they had become so numerous that carbonate oozes began to accumulate on ocean floors in many regions of the Earth. These changed into chalk, such as the famous white chalks of the British Isles and northern Europe. It was for these chalks that the Cretaceous Period in Earth history was named. The surface waters of the oceans probably began to produce a much greater amount of plant food during Cretaceous time.

Many of the marine fishes inhabiting North America after the 94-million-year extinctions had ranges that extended into western Europe and the lands bordering the Mediterranean Sea. This is scarcely surprising, for the British Isles were separated from North America by only 1000 kilometres of water, roughly the combined breadth of the Great Lakes. The continental shelves of the two land masses were but 400 kilometres apart, a distance equal to the length of Lake Huron. But some of the larger fishes (such as the protosphyraenids resembling archaic swordfish, the pachyrhizodontids with trout heads and tuna tails, and the great ichthyodectids) have been found in marine sediments as far away as Australia and Chile. The food chain that sustained them stemmed from microscopic plants pullulating near the surface of the sea. These fishes followed gossamer streams of algal cells to become denizens of a global sea. The expansion of chalk algae with the beginning of the Cretaceous Period may have permitted pelagic fishes to reach the most remote regions of the open oceans, perhaps for the first time. As a consequence, marine backboned animals came to resemble each other around the world, while animals living on each of the separating continental blocks became ever more distinctive.

In North America, however, skeletons of some of the fishes that survived the 94-million-year extinctions are rarely found, or knowledge of their existence is based so far on skeletons discovered in strata of the same age in Europe, Africa, or Asia. This scarcity is at least in part the result of inadequate collecting in North American strata. Later a general planetary warming spilled tropical waters from the Gulf of Mexico across the interior sea in the United States, and far to the north into Canada. With them came the missing fishes and an abundance of marine life derived from equatorial oceans.

The Kansas Chalk

The Niobrara Formation (200 metres thick) is a layer of soft, yellow-weathering chalk that extends in a 2000-kilometre arc from southwestern Kansas to south-central Manitoba.

Niobrara Chalk, exposed near Smoky Hill River, western Kansas

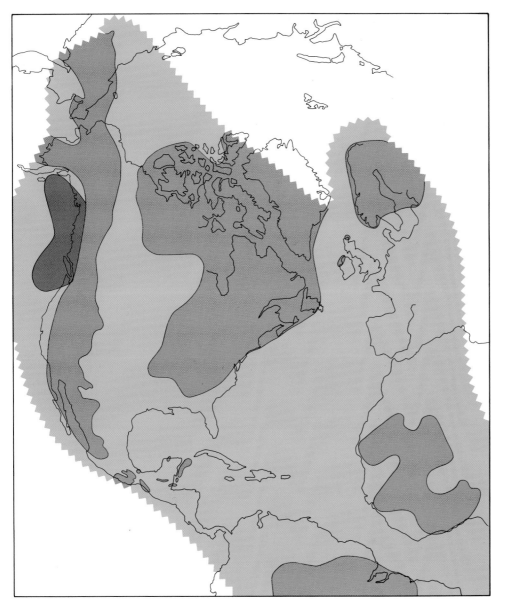

85 million years ago

North America and North Atlantic region. Chalk is being deposited in western Kansas; island terrains are broadening contact with northwestern Pacific coast.

It is best exposed in northwestern Kansas, where badlands have been cut into the chalk along the bluffs of the Smoky Hill River and its tributaries. Spectacularly sculpted chalk pillars may be seen at Castle Rock and Monument Rocks in Gove County. These badlands constitute one of the most important fossil fields in the western United States, and one of the jewels of the geological record of North America. Here famous collectors began their careers, and here during the early 1870s were found the first abundant and complete remains of Mesozoic backboned animals from this continent.

When chalk deposition began in western Kansas 88 million years ago, climates were unusually warm and sea-sonal changes in temperature were negligible. Far to the east lay the low Appalachian peninsula which projected southwest from the Canadian Shield in a manner similar to the way Florida projects southeast of a much more rugged Appalachian region today. Little water flowed into the interior sea from the semi-arid lands to the east. Far to the west lay rugged terrain paralleling the modern Rocky Mountain front, which forced moist air from the sea to rise and release abundant rainfall. Westerly winds from the mountains in turn carried over a hundred volcanic ash falls over the inland sea. Deltaic fans projected into the western margin of the sea, and marine waters were often locally diluted by freshwater sheets. Logs were swept into the sea

Exposed Niobrara Chalk, Smoky Hill River area,
Kansas
—————————
High clouds, near west coast,
Java, Indonesia

by these floods. They were attacked by shipworms, and some sank in Kansas, where they were covered with chalk and turned into pillars of coal.

More impressive than the logs were the coccolith plants of the Kansas chalk. A coccolith platelet measures but 0.005 millimetre in diameter. The coccolith cell, a former member of the group of golden-brown algae, was surrounded by a hollow sphere of platelets measuring 0.01 millimetre in diameter. In the chalk, at what was once the bottom of the sea floor, about half of all of the coccolith platelets seen are packed together in tiny fecal pellets measuring about 0.12 millimetre in diameter. Some of the loose platelets were shed in cell division, or for other reasons, so that more than half of the once-living algae had been eaten by tiny free-floating animals that were nevertheless more than ten times larger than the alga. Life was cheap in the ocean.

The food thus passed along two main streams. One stream was initially large, but the amount of food energy it contained dwindled from alga to small free-floating grazer to small fish to larger fish and finally to a few large carnivores which consumed a small percentage of the original food supply. It was contained in large but relatively rare (fish) parcels. As food passed along the first stream, organic material was lost to a second stream, in the form of platelets, small fecal pellets, large fecal pellets, and finally carcasses that slowly sank to the floor of the sea. This was the staff of life for the detritus feeders.

Chalk algae could not grow well near the western shoreline, because the salinity was lowered by freshwater runoff and the waters were darkened with clouds of sediment. Large marine vertebrates were unable to find food in this aquatic 'desert,' and their skeletons are rarely found here. To the east, as the water cleared, salinities approached normal oceanic levels, and upwelling bottom waters brought nutrients to the surface, chalk algae flourished. Their platelets accumulated on the sea floor at the same rate that particles of silt and mud accumulated further west. The Kansas chalk is essentially a great deposit of tiny coccolith platelets, 'salted' (up to nearly 1 per cent of its volume) with small fish bones. There was food in abundance, and skeletal remains of larger backboned animals occur much more frequently here than on the western floor of the sea.

As mountains rose in the west the chalks were increasingly diluted with sediment and gradually gave way to shale. Large marine animals continued to flourish, perhaps because of the maintenance of the food chain by upwelling bottom waters. However, for over 10 million years chalks continued to accumulate in Kansas, hundreds of kilometres away from the nearest land. The chalk fossils belonged to organisms that lived in an inland sea. They nevertheless provide the best available insight into what life was like in a productive, open ocean during late Cretaceous time.

Sea water is water that is lived in. But organic detritus, avoided by some discriminating organisms and sought out by others, is not without interest. Large fecal pellets, or coprolites, are common in the chalk, and they usually contain fish bones. Most of the food was thus passing from smaller fish to larger fish. But not all. Some coprolites contain the internal supporting structures of very large squids (*Tusoteuthis*) which may have measured several metres in length. Very probably these squids also ate small fish. A few coprolites contain the coiled shells of ammonites, an extinct group of molluscs distantly related to squids. The food requirements of ammonites are not known, but it is thought that they fed on small floating organisms. Squid and ammonite shells are rarely found in the chalk, and the animals may not have been very abundant. An odd accumulation of organic debris occasionally occurs in the form of as many as 1200 intertwined individuals of a free-swimming variety of sea-lily (*Uintacrinus*), distantly related to starfish. These animals evidently floated in swarms, rising and falling in the water in response to the daily movements of the floating micro-organisms on which they fed. The reason for the apparently sudden death of colonies, in which juvenile animals were often abundant, is obscure.

The floor of the chalk sea lay between 150 and over 300 metres below its surface (by comparison, Hudson Bay and the Baltic Sea today are about 200 metres deep, and the North Sea varies between 200 and 500 metres in depth). It was composed of a very soft and watery slurry of unconsolidated chalk, into which carcasses sank. Such bottom currents as existed were very slight, for there is little evidence of surface scouring across an exceedingly flat submarine plain. Volcanic ash falling into the sea formed thin layers on the floor, which were undisturbed by burrowing organisms. The lack of burrowing detritus feeders and the dark colour of fresh, unweathered chalk indicate that oxygen levels were very low. It was an utterly dark and extremely monotonous place, possessing all the gusto and dynamism of a Fifth Dynasty Egyptian tomb.

A constant rain of plant debris, fecal pellets, and tiny floating grazers (all of which was food) wafted sluggishly over the bottom, and there was enough oxygen to sustain organisms possessing either highly efficient gills or low metabolic rates. The waters were clear, so that filter-feeder systems could work. The problem for bottom-dwelling organisms was to be able to pump water through their filter feeders without sinking into the chalky slurry. This was accomplished by giant bivalves, called inoceramids, through two basic adaptations. In some the lower shell became very deep, so that the animal floated in the slurry in a 'boat' and the flat upper shell lay in clear water. In others, a 'snow-shoe' strategy was adopted, so that the thin shells were nearly 2 metres in diameter. The bivalves were distantly related to oysters and have been found to contain fossilized pearls measuring 2.5 centimetres in diameter.

The inoceramids were effectively rafts floating in a sea of ooze, and the rate of platelet fall was slow enough (about 0.035 millimetre per year) that it did not accumulate on their upper shells. Spat settled on them and grew into small oysters which filtered water from the top of an inoceramid island. Usually the oysters formed several layers, and in some instances the topmost layer consists of tiny oyster spat instead of mature oysters. The inoceramid had been over-loaded and sank into the ooze, carrying its load of oysters to perdition. There is a puzzling association between ino-ceramids and small fish, as many as one hundred of which have been found on a single shell. Some fish are preserved enshrouded in the faint, ribbon-like outlines of the bivalve's gills. Small fish have also been preserved lying on the skulls of giant fish and on areas where the chalk substrate was relatively firm. Were the small fish ecologically linked to the inoceramids? Did they aerate their gills in the oxygen-poor bottom waters at the expense of the gills of the ino-ceramid? If the association was the result of the inoceramid having supplied a firm platform on which the tiny carcasses could decay without their bones being disturbed, how did the dead fish enter the inoceramid? Inoceramids could be eaten. In other chalk seas a large, pavement-toothed shark (*Ptychodus*) crimped the margins of the shells or shattered them completely in order to extract the bivalve's soft tis-sues. The shark was present in Kansas and could have descended into oxygen-poor bottom waters long enough to browse on the inoceramids. A few coprolites have been found in the Kansas chalk that do contain broken pieces of inoceramid shell.

The small amount of oxygen in bottom waters was a serious problem for bottom-dwelling organisms, and when circulation diminished for relatively long periods of time these animals temporarily died out. The rain of organic material from surface waters continued, however, and was incorporated into the chalky oozes. This organic material was not recycled, and a long-term leak occurred in the ancient marine ecosystem. Three per cent of the Kansas chalk consists of organic carbon that is still decomposing. Bacterial decay processes have created reserves of natural gas in the chalk that are being exploited commercially. After

Wistari Reef, Great Barrier Reef, Queensland, Australia

many tens of millions of years, organic material formed in a Cretaceous sea is only now re-entering the biosphere. Cretaceous life is still decaying.

Fishing in the Kansas Sea

The carcasses that fell to the sea floor belonged to animals that inhabited the surface waters where most, by far, of the biological activity was taking place. As in apartment build-ings, the parties seem to have taken place overhead, in higher levels of the sea. If the skeletons that resulted from these activities exceeded a length of about 30 centimetres, they were less likely to disintegrate, and as a consequence the larger backboned animals are rather well known. Over 7000 skeletal fragments of swimming animals have been collected from the Niobrara Chalk in Kansas, making it one of the most productive geological units in the world for fossil vertebrates. Far more skeletal fragments have been collect-ed from this chalk than from all of the world's known dino-saur-bearing localities. Nearly 60 per cent of these skeletal fragments are from fish. A blue-water fisherman trolling in the chalk sea could anticipate some interesting catches. Among the more common varieties he might gaff are the following.

Sharks: These are known primarily from teeth, for their cartilaginous skeletons are rarely preserved. In the case of the cretoxyrhinids, relatively complete skeletons have been collected from the chalk, indicating that these free-swim-ming sharks attained lengths of 6.7 metres and weights of about 1.8 metric tonnes. The animals were thus bigger than any securely established record for the modern great white shark. The cretoxyrhinids were formidable pelagic carnivores.

The mysterious shellfish-eating ptychodontids are known only from their heavy teeth, which were articulated together into cobblestone-like crushing surfaces. In one case tiny ossicles imbedded in the skin were found associated with a set of clam-crushers. The mouth was about 35 centimetres wide, in keeping with a shark about 2.5 metres long. Its body was probably intermediate in shape between those of skates and free-swimming sharks. They were not common in the chalk sea and would not ordinarily be caught.

Archaic bony fishes: Pycnodontids were thin, deep-bodied fishes that resembled the modern reef-dwelling angelfish in body outline. The beanlike teeth were set in rows within a narrow pointed mouth and seem well suited for crushing crustaceans. Chalk pycnodonts evidently approached a metre in body length. They may have been caught by using large shrimps as bait.

The billfishes of the chalk sea were the protosphyraenids, of which only skulls and the long, scimitar-shaped forefins are preserved. As in sturgeons, the bones of the body seem not to have been well ossified, but such bony structures as are available are consistent with body lengths on the order of 3.5 to 4 metres. The forefins bear a serrated edge in front and were evidently used to hack schooling fishes to pieces as the large fish swam through them. It would then wheel to consume its stunned and broken prey.

Primitive bony fishes: Related to the basal stock from which mooneyes, herring, and ladyfish descended, ichthyodectids were the dinosaurs among bony fishes. *Xiphactinus* was one of the largest bony fishes of all time, attaining lengths equal to those of the largest living billfish (4.8 metres) and weights of the order of 750 kilograms. Their bodies resembled those of modern tarpon, and their skulls were built like a piscine bulldog, with a powerful, well-muscled crest on the back of the head. Bones of a variety of other fishes are often found within in the stomach region. In some cases undigested skeletons of 2-metre-long fish have been preserved between the ribs of *Xiphactinus* specimens. The larger animals had evidently died of gluttony.

Related to the ichthyodectids were the smaller saurocephalids. Their bodies were garlike in outline, and they attained lengths of 1.2 metres. The lower jaw projected upward at an angle and extended even further in front of the fish's skull than does the jaw of a barracuda. Perhaps they rose beneath their victims, stunning them with a blow from the lower jaw immediately before seizing them. It would not have been difficult to hook ichthyodectids or saurocephalids. These large fishes were structurally rather primitive, and it is tempting to suspect that they lacked the high metabolic rates of some modern big-game fishes. They may not have been hard fighters. The remotely related tarpon is, however, known for its explosive battles when hooked.

Plethodids were the sailfish of the chalk seas. Unlike modern sailfishes they had no bill with which to club smaller fishes but fed instead on thin-shelled floating invertebrates. These they crushed between elongated pads of denticles on their jaws. Plethodids grew to lengths of 1.8 metres.

Among the powerfully swimming carnivorous fishes of intermediate size (up to 1.8 metres) were the pachyrhizodontids, with their wide, lunate tails and troutlike heads. There were several species of differing sizes, although the fish were not common in Kansas seas.

The typical crossognathid was a small, herring-like fish (*Apsopelix*) that fed on floating organisms. It preferred open marine environments but grew to only 35 centimetres in length. Probably an attractive item in the diet of large fish, it may have constituted excellent bait.

Fishes derived from ancestral salmonids: Rivalling the ichthyodectids in abundance, and together with them dominating the fishes of the chalk sea in Kansas, were the enchodontids. There were the 'Tasmanian devil' fish with strong fangs in their oversize heads. Many were not large fish, but several species coexisted in the chalk sea, and some did attain lengths of 1.8 metres.

Cimolichthyids were relatively large fishes, with long, low skulls and elongate bodies. They were powerful, oceanic fishes up to 1.4 metres long. Their smaller cousins, the dercietids, were on average less than half as long, but their bodies were so slender that they were thin even by the standards of modern needlefish.

Spiny-rayed fishes: On land, plants that are continually being browsed often develop thorns. During middle Cretaceous time some small fishes, derived from primitive salmon stock, were continually being eaten by larger fish. They developed spines in front of their fins that could inflict wounds in the jaws of their attackers. The paired fins in the pelvic area were displaced forward, just beneath the chin, enabling the fish to halt suddenly or alter their direction of movement and thus avoid capture. Their jaws were hinged so that they could be quickly projected forward to seize their prey before it could escape. These successful creatures gave rise to the great majority of modern fishes. The tiny, 10-centimetre-long holocentrids of the Kansas sea were the forerunners of most of the marvellous piscine beauty and variety now gracing the world's oceans. Holocentrids, or squirrelfish, are still alive today. They are active, brightly coloured fishes with large eyes and tend to be nocturnal in their habits. The most common fish skeletons found in inoceramid shells belonged to holocentrids. The little fishes were probably much more common in the chalk seas than their rather infrequently found skeletons suggest.

Taken as a whole, the fishes of the chalk seas were a rather odd lot by modern standards. Some of their pecularities as a group were due to the fact that they were living in the open ocean. Hence there were few skates and rays, or bottom-feeding sharks, which prefer muddy, shallow-water environments close to land. Similarly, there were few fishes adapted to exploring the holes and crannies of reefs, for there were no reefs near Kansas. The chalk seas were dominated by huge fishes that were adapted to engulfing smaller fishes, which were in turn adapted to engulfing smaller fishes, etc. Accepting their ecological role as active predators on the high seas, one is nevertheless struck with how elongated their bodies tended to be. Most belonged to the mooneye-herring-ladyfish ancestral stock, the descendants of which today often have transparent, ribbon-like larvae. After these larvae reach a certain size they shrink and metamorphose to resemble miniature adults. Perhaps the chalk carnivores also went through a ribbon larval stage, but the transformation to the adult form was less profound. Modern predatory fishes, such as tuna, possess a teardrop body shape in keeping with the hydrodynamic efficiency required to travel great distances at relatively high speeds. By inference, their counterparts in the chalk sea did not possess this ability to such a high degree.

Large fishes were numerous in the chalk sea, but the number of different varieties seems low by modern standards. No more than about a dozen bony fishes grew to lengths of 1 metre or greater. Off the Atlantic coast of North America today, more than twice as many varieties attain these lengths. It is not quite fair to judge fishes that lived 80 million years ago by modern standards. They had not benefited from a correspondingly long period of natural selection and are thus representative of a less biologically demanding stage of evolution. They were less exquisitely adapted to their niches, there were fewer biological niches for them to become adapted to, and in all probability they were less capable of behaviourally complex activity. It is remarkable that tropical fishes today can so quickly learn to distinguish between a scuba diver, and a scuba diver with a spear gun. Few underwater hunters, however, would have cared to try to ward off a *Xiphactinus* by waving an impressive-looking spear gun.

Air-Breathing Predators

At many different times in the geological past, terrestrial animals abandoned the ecological accomplishments of their remote ancestors, the air-breathing fishes, to reinvade the sea. Ichthyosaurs, plesiosaurs, crocodiles, turtles, snakes, penguins, whales, and seals, to cite a few obvious groups, have established themselves very well in marine environments. In doing so they have mimicked the shapes of fish. However, fish have effectively invaded the land only once, 365 million years ago, and these were the ancestors of all succeeding land-dwelling backboned animals. No fish has since emerged from the water to mimic the shapes of the dominant backboned animals on land. How is it that land-dwelling creatures, though initially adapted imperfectly to living in water, can successfully compete with marine creatures whose ancestors have never left the sea, but that the reverse process never occurs? This is a rather large question. Perhaps an answer lies in the possibility that by breathing air, which is much richer in oxygen than water, an animal can more easily sustain high activity levels which promote survival much more effectively than does a streamlined body shape alone. The most productive waters of the sea occur at its surface, within easy reach of air-breathing animals.

Thus, over 90 million years ago, predaceous lizards closely related to modern monitor lizards invaded shallow marine waters to occupy the ecological niche left vacant by the extinct ichthyosaurs. A few million years later they had become the top predators of the chalk sea and were threatened only by the relatively rare giant sharks. Fully 25 per cent of all skeletal fragments collected in the chalk belonged to the predominantly fish-eating mosasaurs. They are sometimes so exquisitely preserved that delicate structures (such as the cartilaginous eardrum and rings of the trachea, and the tiny bony plates that strengthened the wall of the eye) can easily be identified. The legs of these lizards had been modified into webbed or paddle-shaped flippers, and their tails were spatulate structures like the tails of eels. They swam with undulating movements of the body and tail, steering with their flippers. In some cases the diamond-shaped scales that covered their bodies are still preserved in chalk.

One can visualize a mosasaur seizing a fish and gulping it down by suddenly thrusting its neck forward in the manner of its ancestors, so that the inertia of the fish propels it down the throat. Did they sometimes lift their heads above the water for a gravitational assist? Many coiled shells of ammonites have been found that were punctured by the teeth of mosasaurs. One shell shows in quasi-cinematographic fashion the effects of a mosasaur attack. The shell was first struck several blows in scissors fashion and was then alternately rasped and crushed. Mosasaurs would have seemed like mindless, savage creatures. Their flippers and tails often bear scars from the attacks of other members of their kind.

In spite of the existence of many superbly preserved skeletons, no eggs or young have every been identified within the body cavity of a mosasaur. Hatchlings are very rarely found, although on one occasion the jaw of a tiny mosasaur was mistaken for that of a toothed bird. The resemblance is surprisingly close. Mosasaurs were marine animals, and their remains very seldom occur in estuaries, and never in freshwater deposits. How and where they bred remains unknown, although skeletal remains of juvenile mosasaurs are very common in marine sediments deposited along the eastern coast of the Gulf of Mexico. Perhaps, like some living reptiles, there they sought the secluded beaches of isolated islets and atolls in which to lay their eggs.

There were three major varieties of chalk mosasaurs. *Clidastes*, with a length of 4 metres, was the most lizard-like of the three, but it bore an incipiently symmetrical fin on its tail. *Platecarpus* was 6 metres long and represented something of a cross between a lizard and a seal. It seems to have been exceptionally manoeuvrable. *Tylosaurus* reached lengths of 8 metres, and its jaws were lined with unusually heavy teeth. The muzzle was drawn into a blunt, ramlike prow, which it may have used to stun its prey. On at least one occasion it had indeed been used as a ram, for it was damaged in life and had healed in the shape of a mushroom!

Remains of batwinged flying reptiles of the dinosaurian era are nowhere preserved in greater abundance than in the Kansas chalk. Well over 800 skeletal fragments have been recovered, amounting to 12 per cent of all those of backboned animals collected. Most of these are of very poor quality: their hollow bones are badly crushed, and the skeletal fragments are usually very incomplete. The body of a pterosaur would approximately fill the mouth of a mosasaur, large shark, or giant ichthyodectid, and its bony head and wing membrane were relatively much less appetizing. Such would have been the hazards of alighting on the Kansas sea. The chalk pterosaurs all lacked teeth, but small bones associated with their skeletons suggest that they subsisted on fish. They probably skimmed the surface of the sea, snatching up their prey in their bills and transferring them

to a throat pouch, as do pelicans. Unlike pelicans, their very light skeletons prohibited them from diving into the water.

Filamentous structures associated with the bodies of pterosaurs in fine-grained strata in Europe suggest that their bodies may have been covered with hair. The presence of bodily insulation would in turn suggest high body temperatures and high metabolic rates. The size of the brain cavity is only slightly larger than in an average reptile of similar weight, and pterosaurs were less neurologically sophisticated than birds. Unlike in bats, the wing was supported by only one wing finger, the fourth, but it was further strengthened by cartilaginous rods extending from the leading to the trailing edge. A variety of pterosaurs have been described from the chalk. There were two small forms: those with wing-spans of up to 3 metres lacking any crest on the back of the skull (*Nyctosaurus*), and those with a small crest (*Occidentalia*). The most abundant pterosaur, *Pteranodon*, had a wing-span of 7 metres and possessed a long, graceful cranial crest. A rare form known only from a single skull (*Sternbergia*) bore an enormous crest resembling a pharaoh's crown. The size of the head suggests a wing-span of over 9 metres, making it one of the largest pterosaurs known.

The flight characteristics of pterosaurs have been studied in detail, on the assumption that the atmosphere during the dinosaurian era differed in no aerodynamically important way from that of the present (for example, in mass or density at sea level). This assumption has not been securely established, but the carefully calculated results are nonetheless quite interesting. They indicate that pterosaurs were better gliders than are most soaring birds. The reptiles were superbly adapted for soaring in calm climates, but birds are better fliers under turbulent conditions. *Nyctosaurus*, weighing 2 kilograms, with a wing-span of 2.7 metres, could take off from the surface of the water under its own power and cruised at a speed of 27 kilometres per hour, alternately flapping and gliding. *Pteranodon*, weighing 15 kilograms, with a wing-span of 7 metres, could not rise from the sea without the help of a gentle breeze and soared at a velocity of 31 kilometres per hour. Its crest gave it no special aerodynamic advantage, for many pterosaurs lacked the structure. From western Kansas the nearest shoreline to the west was 720 kilometres away, and to the east the distance was 500 kilometres. Hence, under windless conditions, *Nyctosaurus* was 18.5 hours and *Pteranodon* 16 hours from land. The animals could have flown against trade winds, but stronger winds could have been avoided only by landing on the water. Skeletal fragments suggest that this option was sometimes detrimental to a pterosaur's health.

Birds were present, both over and within the chalk sea but they were figuratively, and perhaps also literally, overshadowed by pterosaurs. About 3 per cent of the skeletal fragments in the chalk are avian, although this figure is probably high, as the remains of birds with teeth were greatly prized by collectors. There were two major varieties, and in both the bills were indeed partly lined with small teeth. Perhaps the most famous are the hesperornithiforms. The skeletons of these flightless swimmers were made of very heavy bone in order to give the otherwise buoyant birds sufficient ballast to equal approximately the density of water. Their robust bones were easily preserved in the geological record. The second variety are the ichthyornithids, in which are included the genus *Ichthyornis* with many distinct species. The bones of these birds were not heavily constructed, and although they are as abundantly preserved as are those of hesperornithids, *Ichthyornis* may well have been even more common, because its bones were exposed to a greater risk of destruction. These birds were gull-like and must have been powerful fliers, for they were able to range as far out to sea as were the pterosaurs.

Hesperornithiforms were the small, warm-blooded pursuit predators of the seas. Their bodies were covered with hairlike feathers, and their lobed feet were scaled. They propelled themselves through the water with kicks of their hind legs somewhat in the manner of giant diving beetles. Their legs could not be brought under their bodies to stand on land. In an additional and fundamental difference from penguins, which 'fly' through the water, the wings of hesperornithiforms were very small, and most probably served as rudders, like the lateral fins of squids. Their necks and skulls were generally long and loonlike, and coprolites associated with skeletons bear witness to an appetite for small fish. Three different varieties are known: *Parahesperornis*, with a body length of under 1 metre, *Baptornis*, which exceeded 1 metre in length, and *Hesperornis*, which grew to nearly 2 metres. About one-third of the *Baptornis* specimens were of juveniles, and the animals must have nested along a nearby coast. None of the numerous specimens of *Hesperornis* from the chalk was immature. The presence of such small and energetic animals confirms the high productivity of the Kansas sea. These birds had no defence other than their agility, and their success does not speak highly of the pursuit skills of their larger contemporaries.

Remains of marine turtles are about as abundant as those of birds, comprising 3 per cent of the specimens of backboned animals that have been recovered. In spite of the paucity of fossils, at least 10 varieties of marine turtles have been identified. The turtles were fully marine and 'flew' through the water by flapping their foreflippers. Some sprinted by kicking with their hind flippers as well. What constituted turtle food is conjectural, but in an ecosystem dominated by fishes it might be suspected that fish-eaters would also be common. In view of the scarcity of their remains, the turtles more probably fed on bottom-dwelling organisms in shallower waters, including kelp, and were only sporadic visitors to the chalk sea. The most common turtle fossil is a skull, although the shells typically measured between 0.5 and 1.5 metres in diameter. A few bore conical 'thorns' along the midline of the upper surface of the shell, evidently for defensive purposes. These have been termed 'sawturtles.' One is left with the image of a *Tylosaurus* or a shark crushing a turtle shell to bits and swallowing the pieces, while the head becomes detached and falls to the sea floor. Turtle remains are so abundantly preserved in marine deposits from Arkansas to Alabama that the region was the 'Tortuga Coast' of late Cretaceous time.

Only 1 per cent of all the skeletal fragments of back-boned animals belonged to plesiosaurs. The ecological roles of the two major plesiosaur groups are not well understood. Did the large-headed polycotylids dive in pursuit of large squids in the manner of sperm whales? Was the rareness of the long-necked elasmosaurs due to their preference for bottom-dwelling fishes? Their skeletons are so uncommon in the chalk that those that have been preserved may have belonged to animals that were only migrating through the midcontinental sea. Skeletal parts of very immature plesiosaurs have been collected from the chalk.

Dinosaur skeletons are the rarest of all. Only four have so far been reported, and like the carbonized logs in the chalk the arrival of a very few bloated dinosaur carcasses in western Kansas may be considered accidental. One skeleton was of a 5-metre-long juvenile duck-billed dinosaur, the only known specimen of *Claosaurus*. The animal had a short shin, as did flatheaded duck-bills, and was primitive by later standards in still possessing a small 'big' toe. All of the remaining three skeletal fragments belonged to an armoured dinosaur called *Hierosaurus*, a variety of nodosaur. It will be recalled that nodosaurs bore heavy spines on their necks and shoulders and possessed a long tail protected by scutes but lacking a terminal tail club. Its unusual relative abundance suggests strongly that the animal preferred a habitat close to the coastline from which carcasses would have an unusually high probability of being washed out to sea.

To the Edges of the Kansas Sea

The Niobrara Formation extends north through Nebraska and the Dakotas into Manitoba, lapping around ancient islands in southwestern Minnesota. These islands were as far from western Kansas as western Kansas was from the western and eastern shores of the sea. It would be interesting to know if they were an important haven for air-breathing marine animals. The chalk is poorly exposed on the northern plains, and few fossils have been collected from it. A flipper of the mosasaur *Platecarpus* was found in North Dakota. Roadcuts into the chalk in Riding Mountain National Park, on the Manitoba escarpment, have produced bones of the mollusc-eating shark *Ptychodus*, giant ichthyodectids, and other fishes that lived in Kansas seas, as well as of the diving bird *Hesperornis*. To the west, isolated specimens of mosasaurs (*Platecarpus, Tylosaurus*) have been collected from chalk-age sediments in Montana, Wyoming, Colorado, and New Mexico. The few available fossils give no evidence of an ecology different from that of western Kansas.

In northeastern Texas, a little over 100 kilometres northeast of Dallas, thin limestones (Ozan Formation, 4 metres thick) show the effects of strong bottom currents. The most common mosasaur bones found here belonged to the giant *Tylosaurus*, which in the warm seas of Texas apparently became twice as long as in western Kansas. Here, it attained lengths of 16 metres. Unlike in western Kansas, spherical teeth of the shellfish-eating mosasaur *Globidens* have been discovered, indicating that the currents brought well-aerated waters to the bottom of the shallow sea and supported a richer bottom fauna of shellfish on which the mosasaur grazed. Some 400 kilometres to the east, and 30 kilometres south of Little Rock, Arkansas, lime-rich sediments (Marlbrook Marl, 65 metres thick) have yielded a few fossil specimens of backboned animals. Among these are one of a mosasaur (*Platecarpus*) and another of a plesiosaur with a neck of intermediate length (*Cimoliasaurus*). The remaining six specimens belong to four different varieties of marine turtles, a very unusual circumstance considering the rarity of turtles in western Kansas.

In central Alabama, chalks (Mooreville Formation, 75 metres thick) are exposed sporadically in ravines in an arc extending southeast to Montgomery and east toward Georgia. These chalks contain remains of many of the same varieties of backboned animals that inhabited the Kansas sea, but there were profound ecological differences between the two regions. Turtles make up fully 40 per cent of the specimens of backboned animals collected here. Their abundance does not seem to affect the ratio of mosasaur to

Islands in chalky sea,
Florida Keys

fish remains, again suggesting that turtles were not competing with mosasaurs for piscine prey. The ram-nosed mosasaur *Tylosaurus* is most abundant at the base of the chalk, and its smaller relatives *Platecarpus* and especially *Clidastes* replace it in higher levels, where remains of the ball-toothed mosasaur *Globidens* occur as well. Pterosaurs were not common in Alabama skies, and no remains of *Hesperornis* have ever been found.

As in the Kansas chalk, teeth belonging to bottom-dwelling sharks are rare, and open marine conditions prevailed. However, in the Alabama chalk the tiny denticles have been found that were once imbedded in the skins of whale sharks (*Rhiniodon*). Here, in subtropical waters off the southern coast of the United States of 80 million years ago, lived the most ancient known large vertebrates that were able to feed on tiny floating organisms strained from sea water. In doing so they were able to capture a much larger proportion of the available food than would predatory animals feeding on large fishes. They anticipated the ecological role of baleen whales by several tens of millions of years. The giant ichthyodectids were less abundant than

in Kansas, and the fauna was dominated by relatively smaller fishes, such as the pachyrhizodontid 'trout-tunas' and ram-jawed saurocephalids.

A few remains of dinosaurs have been found in the Alabama chalk. One bone was from the toe of a bipedal carnivore, another from the pelvis of a juvenile armoured nodosaur. A partial skeleton of an immature flatheaded duck-billed dinosaur (*Lophorothon*) about the size of the specimen from Kansas has also been collected. It differs from its western relative in having a longer shin and differently constructed teeth. There is a large gap between the bones on top of the head, as in human babies. The Alabama duck-bill may have differed from the Kansan dinosaur because it was from the Appalachian region, while the Kansas specimen washed from the western subcontinent across the inland sea. But what were the environmental conditions that accounted for the difference in the abundances of other marine animals between Alabama and Kansas? Perhaps the turtles were commuting to and from nesting grounds on offshore sand islands. The Alabama chalk was certainly much closer to land. However, it may not have received the flux of inorganic nutrients that poured into the interior sea from deltas bordering the western mountains. Do large plankton-feeders, such as whale sharks and baleen whales, diminish the stock of small food organisms so greatly that the number and size of predatory fishes are equally diminished?

Near the western border of Georgia the chalks are replaced by sands (Blufftown Formation, 180 metres thick), and in turn by sediments deposited in near-shore lagoons. The latter are exposed along streams in North Carolina (Black Creek Formation, 120 metres thick) and in Delaware and New Jersey (Merchantville and Woodbury formations, totalling 33 metres in thickness), in a narrow band paralleling the modern coastal plain to the northeast. The lagoonal sediments are composed of clays, silts, and fine-grained sands located near the delta fans of rivers and often contain carbonized logs and plant debris. Many of the large predators that inhabited the Kansas sea, such as the bony fishes *Enchodus* and *Xiphactinus* and the mosasaurs *Clidastes, Globidens, Platecarpus*, and *Tylosaurus*, frequented or at least were washed into these lagoons after death. The skeletons of all backboned animals were disarticulated, perhaps by scavengers and perhaps by tidal currents, so that no even partially complete skeleton has been recovered. Nevertheless, available remains indicate that marine life along the Atlantic coast differed from that in the chalk seas.

The teeth of cartilaginous fishes are preserved abundantly in Atlantic strata and can easily be identified. Coastal waters are thus known to have been inhabited by numerous bottom-dwelling sharklike fishes, such as angel sharks, sawfish, guitarfish, skates, and rays. Powerful predatory sharks were present as well. Among the bony fishes, those forms (such as pycnodonts) feeding on shelled invertebrates were more common than in the interior or southern seas. The pavement-toothed shark *Ptychodus* is, however, not known from the region. Neither have any remains of the flightless diving bird *Hesperornis* been identified. Marine turtles were present but were apparently less numerous than along the 'Tortuga Coast.' Because of the near-shore origin of the sediments, dinosaur bones are relatively common, but they too are completely disarticulated and usually water-worn. Perhaps the most spectacular member of the Atlantic community was a gigantic true crocodile, named *Deinosuchus*, that apparently attained a length of 15 metres. The animal was a match for any predatory dinosaur of its day. Although it is also known to have inhabited deltaic environments along the western shore of the interior sea at this time, *Deinosuchus* is more extensively represented in sediments of the Atlantic region.

The Sea in Time

The marine animals of the Kansas sea maintained their identity as a whole for many millions of years. However, just as one group played a more important role in one region than in another, the various groups also changed in relative importance through time. While in the former instance the changes contributed to the geographical interest of various parts of the sea, in the latter case biohistorical or evolutionary changes give depth to a timescape. Most temporal changes were the result of geologically short-term changes in the physical environment. Rugged terrain was being elevated into mountains along the western border of the seaway, while its basin apparently deepened. Chalk deposition was shifted progressively further toward the east, and bottom waters evidently became stagnant and more acidic. On a global basis, mean temperatures declined from an optimum that coincided with the initial period of chalk deposition. The biological response kept abreast of these physical pacemakers. The response was typical of interactions between organisms and various, perhaps unknown physical factors. Over tens and hundreds of millions of years, the effects of interactions between various organisms become more clear-cut and have as their result an increase in organic complexity. The latter effects can be seen by comparing fishes of the ancient Kansas sea with those of the modern Atlantic coast.

As the sea deepened in Kansas, deep-water environments changed, perhaps as a result of falling oxygen concentrations or chilling. The large inoceramids disappeared, together with pavement-toothed sharks (*Ptychodus*) and shellfish-eating pycnodonts and 'sailfish' (plethodids). Relative abundances of predatory fishes inhabiting surface waters apparently shifted for reasons that are not immediately clear but may reflect the quality of algal food at the base of the food chain. Thus the archaic billfish with scimitar fins (protosphyraenids) and 'Tasmanian devil' fish (enchodontids) declined, to be partly replaced by the giant ichthyodectids, 'ram-jawed' fish (saurocephalids), and 'trout-tunas' (pachyrhizodontids). In similar fashion there seems to have been a change in the relative abundance of different mosasaurs, so that the larger *Tylosaurus* was replaced partly by the smaller *Clidastes*. Perhaps turtles became more abundant because they were less often an alternate food for *Tylosaurus*. Plesiosaurs and pterosaurs possibly declined, and toothed birds increased in number. In sum, the average size of predatory fishes increased, while the average size of air-breathing predators diminished. Are most of these changes simply related to an environment that more and more resembled that of deep oceans? Or was a slow but general decline in mean annual temperatures more important? Clues must be sought in the geographical distribution of the animals.

About 79 million years ago, chalk deposition ceased all over the basin of the interior sea. Chalky mudstones were succeeded by black, sulphurous shales which accumulated under anoxic, acidic conditions on the sea floor. Surface waters may not have changed so abruptly, however, for the fauna of the Kansas sea is well represented in sediments deposited across the northern plains (Pierre Formation: Sharon Springs Member, 50 metres thick in Kansas, Wyoming, and South Dakota; and Pembina Member, 25 metres thick in Manitoba). The fossil fishes of these deposits are not well studied, but the large cretoxyrhinid sharks seem not to have been present. *Tylosaurus* continued its decline, but *Clidastes* is less numerous relative to *Platecarpus* remains as well. Both plesiosaurs and toothed diving birds were more abundant than in chalk seas. No sawturtles have been collected, but remains of marine turtles with smooth shells are relatively common. Perhaps the need to discourage the attacks of turtle-eating mosasaurs was less. The relative abundances of most air-breathing animals changed according to a pattern that was probably established during the time of chalk deposition.

Two thousand three hundred kilometres to the north of the US-Canadian border, sulphurous smoke rises throughout the year from cliffs on the edge of the usually frozen Beaufort Sea. The combustion results from the exposure of carbon-rich, oxygen-starved sediments to the air. The dark strata (Smoking Hills Formation, 40 metres thick) are dotted with areas where the clays have been fired to a bright red colour. They are now well north of the Arctic Circle, and the region was even further within the arctic region of 79 million years ago. Representatives of the backboned animals of the Kansas sea have been collected in these exposures, including giant ichthyodectids, mosasaurs, plesiosaurs, and toothed birds. The relative abundances of the remains of the air-breathing animals are more extreme than far to the south. *Platecarpus* is the only known mosasaur, and *Clidastes* and *Tylosaurus* are either rare or absent. Plesiosaur remains are three times as abundant as those of mosasaurs, and nearly half of the specimens of air-breathing vertebrates collected belonged to *Hesperornis*. No remains of pterosaurs or marine turtles have yet been found.

Significantly, there were no forms discovered that were unique to the polar region. The animals seem to have been migrants from the south, and a large proportion of the plesiosaurs and toothed birds were juveniles. Perhaps they were breeding in the far north during summer months. Strata

like those in the cliffs along the Beaufort Sea have been traced 500 kilometres to the north into the Queen Elizabeth Archipelago. There, on Eglinton Island, *Hesperornis* and chalk fishes have been found at a latitude midway between the northernmost tip of Alaska and the present rotational pole. The changes in abundances of marine animals through time in the chalk of western Kansas were evidently similar to the changes that took place in space between western Kansas and the high arctic, though less extreme. Therefore most of the shifts in abundances of various backboned animals in western Kansas were probably not caused by an increasingly oceanic environment. The cause was more likely a slow decline in mean annual temperature that was less extreme than an otherwise similar decline in mean annual temperature between midlatitudes and polar regions.

A Faunal Turnover

About 75 million years ago the continuity of the history of backboned animals of the chalk sea was disrupted. Many of the large chalk fishes apparently vanished from the interior of North America, including the scimitar-finned protosphyraenids, the giant varieties of ichthyodectids, and the ram-jawed saurocephalids. The mosasaurs *Clidastes, Globidens, Platecarpus,* and probably also *Tylosaurus,* which had taken part in the faunal evolution of the chalk sea for more than 10 million years, also vanished. A bulky species of *Platecarpus* appeared 1 or 2 million years before the turnover, but it vanished with the other mosasaurs. No remains of the giant crocodile *Deinosuchus* have been found in younger estuarine environments bordering the interior sea. If the turnover were caused simply by a sudden drop in temperature, it might have been expected that a northern fauna composed of ichthyodectids, *Platecarpus,* plesiosaurs, and *Hesperornis* would have spread far south toward the Gulf of Mexico. Instead, the new fauna was dominated by smaller fishes and the slender, relatively highly evolved mosasaurs known as *Mosasaurus conodon* and *Plioplatecarpus,* which descended probably from southern *Clidastes* and northern *Platecarpus,* respectively.

All the forms that became extinct within the interior of North America did survive in warmer regions, except for the diving birds, which survived at least in Chile. It would appear that while temperature may not have been the only cause of the faunal turnover in the shallow, interior sea, it was somehow involved in allowing the same creatures to survive in other regions. Was the turnover partly the result of a lowering of the level of the world ocean and of the

Hesperornis

LATE CRETACEOUS, ABOUT 80 MILLION YEARS AGO

Hesperornis parents join chick beside Canadian
arctic stream. Flightless marine birds breed, and
stream meander has breached ring of northern conifers.

withdrawal of seas from the North American continent? But marine sediments postdating the turnover have been found near the border between Alberta and the Northwest Territories, suggesting that the marine corridor to the Arctic sea remained open. And identical marine shellfish in South Dakota and western Greenland suggest the existence of a seaway linking the two areas across what is now Hudson Bay. The interior sea continued to cover the plains states. A continent-wide event of some kind caused the faunal turnover, but its nature is conjectural.

Thick sequences of grey-brown shales were deposited on the northern plains after the faunal turnover (upper Pierre Formation in South Dakota, 530 metres thick; Bearpaw Formation in Montana and Saskatchewan, 300 metres thick). The skeletal parts of mosasaurs are the most abundantly found remains of large backboned animals. In addition to the more common *Mosasaurus conodon* and *Plioplatecarpus*, specimens of the more powerfully skulled mosasaurs *Mosasaurus missouriensis* and *Prognathodon* have occasionally been collected. The animals may have preferred shellfish to fish, and thousands of shells of ammonites that were punctured by robust mosasaur teeth in Alberta may be cited as evidence. Plesiosaur remains are about half as common as those of mosasaurs. A specimen of the long-necked elasmosaur *Alzadasaurus* collected in Montana was found to contain nearly 9 kilograms of ballast in the form of river-transported cobblestones. The animal had evidently swum through brackish estuaries to penetrate a river sufficiently far to reach water that was flowing rapidly enough to transport large stones.

Far to the north, on the edge of the Beaufort Sea, grey marine shales were being deposited (Mason River Formation, 150 metres thick). Plesiosaurs again were relatively more abundant than in the south, and their remains are found as often as those of mosasaurs. Most of the mosasaur specimens belonged to *Plioplatecarpus*, and *Mosasaurus* was very rare. The fossils found in the shales of the post-turnover sea do not give the impression of richness, whether in geographical variation, variety of different forms present, or abundance. However, the continuity of the fauna remained intact for some 7 million years. Finally, about 3 million years before the end of the dinosaurian era, the interior sea began to withdraw from its western edge. There, marine shales were replaced by estuarine sands and deltaic muds. The evolution of marine fishes to the end of the dinosaurian era cannot be studied in sediments left behind on the midcontinent.

Sharklike fishes inhabiting the western estuaries resembled their contemporaries on the Atlantic coast, although ratfish (chimaeras) were more common in the west. One group (rhinochimaerids) that today lives only in abyssal depths in the ocean was present in the shallow, brackish waters. The garlike aspidorhynchids, which several tens of millions of years previously had been common in marine environments, were now found, in low abundance, only in brackish environments. Plesiosaur remains are commonly found in ancient tidal channels, and in some cases their bones occur rather commonly in sediments that were deposited in fresh water. Did they enter fresh water to free themselves of marine parasites? In Alberta, a single specimen of 75-centimetre-long ancestral tarpon (*Paratarpon*) was collected in estuarine strata a few million years older. 'Tasmanian devil' fish (enchodontids) and 'needlefish' (dercetids) occasionally entered shallow coastal waters from the sea.

The slowly moving deltaic rivers contained a fish assemblage at once familiar and bizarre. Then, as now, the fishes dwelling in fresh waters belonged to groups different from those inhabiting the seas. Many of these ancient river fishes, such as sturgeons, gars, and bowfins (amiids), would have appeared much as do their living descendants. In comparison with salt-water fishes of their day, these fishes were decidedly more primitive in aspect, for they belonged to the lower bony fish groups, where the skeleton is imperfectly ossified or the body is covered with heavy, armoured scales. But sharklike fishes may well have flourished in fresh waters, in contrast to the present situation. A rayfish (*Myledaphis*) was very abundant in many of these rivers, and shark (*Squatirhina*) teeth that have been assigned to the living wobbegongs ('catfish sharks') seem restricted to fresh waters then. Far away from the coast, living in ponds or small streams, were drums (*Platacodon*), related to the modern perches. The greatest peculiarity of the fishes of the ancient deltas was the rareness of perchlike fish and the complete absence of modern fishes such as minnow, suckers, catfish, basses, sunfish, and sculpins.

The Greensand Sea

Greensand obtains its dark, grey-green colour from a potassium iron silicate called glauconite. The mineral forms in warm, shallow seas that receive a large supply of organic detritus and a relatively small supply of sediment. The organic material is recycled by a teeming fauna of detritus-feeding worms, snails, and clams. Their burrows are often clogged with glauconite. So much organic carbon is mixed with the lean supply of sediment that the bottom sediments become

Greensands in Inversand Marl Pit, near Sewell, New Jersey.
Late Cretaceous greensands (Navesink and basal Hornerstown formations) near
bottom of pit have produced bones of dinosaurs and marine reptiles;
greensands exposed in upper portion (main body of Hornerstown Formation)
were deposited soon after dinosaurs became extinct.

depleted in oxygen, and iron sulphide can form within them. Just such conditions often prevailed along the coast of Maryland, Delaware, and New Jersey during the 10 million years between the faunal turnover and the end of the dinosaurian era.

The shallow marine sediments were deposited in a cyclic manner, in response to cyclic changes in sea level. Thus an offshore greensand was covered by a lagoonal silt, which in turn was covered by a near-shore sand. Sea levels then fell slightly, and the process was repeated. There were three cycles of greensand deposition on the central Atlantic seaboard during the last 10 million years of the dinosaurian era (from oldest to youngest, they are the Marshalltown greensand, Wenonah silt, Mount Laurel sand, Navesink greensand, Redbank silt, Tinton sand, and Hornerstown greensand, with an aggregate thickness of 146 metres). The region became one of the great greensand areas of the globe. A hundred years ago the glauconite was mined extensively as fertilizer, and it is still used as a water softener. However, the commercial uses for glauconite have declined, and only one of the old pits is still being quarried.

Fossil bones extracted from the greensand quarries were among the first specimens collected in North America from dinosaurian-age strata. Because they are permeated with iron sulphide they have a rather pleasant matchlike odour and unfortunately break apart as the chemical decomposes on exposure to air. The skeletons, particularly of the smaller organisms, had usually been scattered before burial. They become even more fragmented through disintegration, following long periods in storage. The specimens nevertheless constitute the best record available in North America of life in the sea during the last 10 million years of the dinosaurs.

The seas along the Atlantic coast were warmer than in the interior, and some of the old chalk fishes, such as the shellfish-nibbling pycnodonts and ram-jawed saurocephalids, were still present. 'Tasmanian devil' fish (enchodontids) inhabited the warm coastal waters, and giant predatory cretoxyrhinid sharks swam off New Jersey shores as they had once cruised the Kansas seas. Sharklike fishes along the Atlantic coast were essentially unaffected by the faunal turnover. Turtles were nearly as abundant as they formerly were along the old 'Tortuga Coast,' and sawturtles were similarly well represented. Aquatic turtles called pelomedusids were quite abundant. Termed 'side-necked turtles,' they now inhabit fresh waters in the Southern Hemisphere. They protect their heads by tucking them to one side under the margin of the shell, rather than by withdrawing them into the shell. As in other warm-water environments, plesiosaurs were relatively rare, although the one rather well-known form (*Cimoliasaurus*) occurs in strata of similar age in arctic Canada. Isolated bones indicate the presence of several different kinds of modern-appearing shorebirds, and high overhead large pterosaurs (*Titanopteryx*) evidently soared (they are known so far only from a single vertebra).

The greensand sea was peculiar for the large number of crocodilians that it supported. It could fairly be termed the 'crocodile coast' of the continent. An ancient branch of crocodilians was represented by a gavial-like marine form, with a long snout and eyes on the side rather than the top of the head. Called dyrosaurs, the animals had rather long forelimbs. They seem to have been most abundant in Africa, where they grew to lengths of 9 metres. The New Jersey representative (*Hyposaurus*) was only about 2.5 metres long. Another, less fully marine form (*Thoracosaurus*) belonged to the same group as do modern true crocodiles. Its skull had a very long slender snout, and the animal probably attained lengths of the order of 8 metres. Crocodilians with narrow snouts, like the above two forms, are usually fish-eaters. The giant, 15-metre-long *Deinosuchus* had a broad snout. Modern crocodilians with broad snouts feed on large animals and tear their prey apart by spinning their bodies in the water with part of their prey clamped between their jaws. The damage a *Deinosuchus* could thereby inflict must have been horrible to behold.

There were as many different kinds of mosasaurs on the Atlantic coast as there once were in western Kansas, but their shapes were no longer the same. Mosasaurs were becoming more perfectly adapted to marine environments; their tails were expanding into fins and their flippers were becoming flukes. The bones on the roof of the skull were beginning to grow over each other toward the back of the head like shingles. The same process has occurred in the evolution of whales. Mosasaurs were becoming the toothed whales of the dinosaurian era. There were several species of *Mosasaurus*, the largest of which probably grew to a length of 13 metres. *Plioplatecarpus* attained a length on the order of 6 metres, and the various other mosasaurs were of intermediate proportions. They all closely resembled contemporaneous species from the chalk seas surrounding the archipelagos of western Europe and may well have been indistinguishable from them.

Skeletal fragments of dinosaurs are seldom found at all in marine deposits, but in New Jersey nearly one-fifth of all specimens recovered belonged to dinosaurs. Very few contain more than one bone, and their body forms remain frustratingly conjectural. Their abundance, however, does indicate that nearby Pennsylvania bore substantial dinosaur populations. Most of the bones belonged to duck-billed dinosaurs, and a very few are from armoured dinosaurs. A variety of medium-sized carnivores seem to be represented by isolated bones, suggesting that the animals often patrolled the edge of the sea in search of stranded marine animals.

Sunsets over the Cretaceous Sea

Between San Francisco and Los Angeles, the El Camino Real winds through grass-covered hills dotted with California oaks. In the spring, when the grass is green and the poppies are blooming near old adobe missions, there are few more beautiful places in the world. When the padres were there, California was a land at the utter end of the Earth, and had they known of giant reptile bones in the hills they would have marvelled but not been too surprised. In ravines west of Fresno, dark shales are exposed (Moreno Formation, 575 metres thick) that were deposited in an arm of the Pacific Ocean when greensands were accumulating on the Atlantic coast.

The fossils are often not as well preserved as specimens from other regions of the continent, but they clearly show the importance of the region in understanding the ancient biogeography of North America. Several specimens of duck-billed dinosaurs have been collected and appear to resemble varieties that inhabited lands to the east, along the western edge of the interior sea. The California land mass, formerly an island in the eastern Pacific, was, therefore, by that time in contact with the rest of the continent. The fishes have not been well studied, but remains of scimitar-finned protosphyraenids, large ichthyodectids, and 'Tasmanian devil' fish (enchodontids) suggest a general similarity to the fishes of eastern waters. There are no indications that they belonged to geographically distinct forms. A turtle known to have inhabited the Atlantic coast during this time (*Osteopygis*) has been identified in the California shales.

The mosasaurs do belong to geographically distinctive forms. *Plotosaurus* was a small animal, ranging between 3.5 and 6 metres in length, with an enlarged tail fin and whale-like skull. It evidently subsisted on small fish. *Plesiotylosaurus* was a larger animal that probably attained lengths of 7.5 metres and possessed heavy teeth suited for crushing large prey. The animals were descended from ancestral forms related to *Clidastes* and *Platecarpus*, respectively. The California mosasaurs were advanced members of a rapidly evolving group. It would be consistent to suspect that their ancestors lived along the coasts of both the Pacific Ocean and the interior sea and then were separated by a geographical barrier. The occurrence of North American land-dwelling dinosaurs in South America at this time, and of South American dinosaurs in North America, suggests that this barrier was a north-south land-bridge linking the two continents.

The California plesiosaurs all belong to the long-necked elasmosaur group. Long-skulled polycotylids have not been recorded here, and their absence (or lack of abundance) may have been a result of either geographical isolation or an unfavourable environment. Most of the elasmosaur skeletons were either from specimens that were immature or from adults measuring less than 9 metres in length. They were smaller than the largest plesiosaurs inhabiting the interior sea. A relatively large number of varieties have been identified (*Aphrosaurus, Fresnosaurus, Morenosaurus, Hydrotherosaurus*), none of which is known outside California. This was another effect of the geographical isolation of the region. Like their contemporaries on the east and west coasts of North America, however, these plesiosaurs participated in the culmination of the great natural experiment of life in the seas during the dinosaurian age. And, as on the land, they were soon to disappear in one of the great natural catastrophes that the planet periodically experiences.

In perspective, the record left behind by Cretaceous seas is unusually rich because of its great extent in time and broad distribution over North America. The continent felt the breath of the ocean, where during Triassic and Jurassic time it had felt the sting of blowing sand. It is well known that life on land during the dinosaurian era was vastly different from what it is now. Life in the water during this time was equally distinct. It was above all less complicated. The bony fishes had not departed greatly from the body plans of their heavily scaled ancestors, and there were far fewer varieties of them. These observations are strikingly relevant to the freshwater fishes. The great adaptive radiation of the higher sharks and spiny-rayed fishes had hardly begun. Many of the fishes were giants, even by modern standards. They seem to have been well dispersed over the oceans, to a greater extent than were air-breathing marine animals. And the fact that identical creatures inhabited both midcontinental and arctic seas indicates that northern environments were not so special as they now are. A unique indigenous fauna was therefore not present in the far north.

Within a broadly interconnected region, it is remarkable how local environments affected the balance between different air-breathing animals in the various subregions. Examples include the relative abundance of pterosaurs in Kansas, of turtles in Alabama, of crocodilians in New Jersey, and of hesperornithiforms in arctic Canada. It is surprising that small flightless marine birds could have ventured so far from land, out into the midcontinental sea in Kansas. Similar changes in the representation of various groups of fishes also existed. It is interesting that a single environmental variable, temperature, may have produced similar changes both from the south to the north and from an earlier to a later period in the life of the interior sea. The record of the Cretaceous marine animals endured long enough to reveal two large-scale phenomena in the history of life. One is exemplified by three abrupt biotic turnovers, increasing in amplitude from the one occurring 75 million years ago, to the one at 94 million years ago, to the planetary catastrophe that terminated the dinosaurian era 65 million years ago. These events are thought to be the result of perturbations in the physical environment; were they all caused by the same mechanism? The other phenomenon may be illustrated by the history of mosasaurs; they were gradually transformed by competition with others of their own kind from lizards into animals that resembled primitive whales in 25 million years. Of the two phenomena, one of extinction and the other of development, the latter has been the more important.

7

Corythosaurus

LATE CRETACEOUS,
ABOUT 75 MILLION YEARS AGO

At dawn, groups of duck-billed dinosaurs (*Corythosaurus*, foreground) and horned dinosaurs (*Chasmosaurus*, background) leave the waters of a bayou, in Dinosaur Provincial Park, southern Alberta. A small carnivore (*Troodon*) pauses beneath a magnolia; ostrich dinosaurs approach water's edge to drink; a soft-shelled turtle (*Aspideretes*) suns itself.

The
Late Cretaceous

A Cretaceous Camelot

During all of late Cretaceous time (between roughly 90 and 65 million years ago) the level of the oceans stood higher than now. The continental margins and broad, slowly subsiding interior troughs were flooded by the sea. Sediments shed from bordering mountain chains and ancient continental shields were carried by streams directly into nearby marine waters, and so for the first half of this period they preferentially preserved an abundant record of marine life. Dinosaur bones sporadically dropped from floating carcasses washed to sea in floods or storms. Only very rarely did skeletal fragments reach the bottom of the sea. But during the latter half of this long period, floodplains and deltas spread from the western shore out toward the centre of the interior sea. In them were buried the bones of land-dwelling creatures. These bones represent the flower of the dinosaurian world, and nowhere else on Earth have dinosaur bones been found in such abundance.

These dinosaurs exhibited a sophistication of form that was not so apparent in their predecessors. Although some were giants, body weights tended on average to be less, bodily contours were more graceful, and limbs were relatively longer. Many had brains that were larger in proportion to their bodies than in modern reptiles. These attributes imply that metabolic rates and activity levels had increased to become transitional to those typical of modern warm-blooded birds and mammals. Accompanying relatively advanced physiologies and behaviours was a differentiation of basic body patterns into a large number of distinctive varieties, adapted to the numerous ecological variations of the lowland environmental theme. It was a world more full of evolutionary excitement and potential than had ever existed before on the planet. The record of lowland life is best sampled in strata representing the last dozen million years of late Cretaceous time. Because of the relatively short span of time, the interplay of ecologies is much more obvious than the effects of further evolutionary achievement.

Dramatis Dinosaurae

Dinosaur faunas of the late Cretaceous of North America were dominated by five major kinds of large dinosaurs. Other dinosaurs were present, but they were usually either much less abundant or much smaller. Foremost among the dominants were bipedal herbivores termed hadrosaurs, a name derived from Greek roots meaning 'heavy lizard.' In English, they are popularly called 'duck-billed dinosaurs,' which has been translated as such into languages as diverse as Chinese and Russian. When shown a reconstruction of a hadrosaur, a man belonging to the Turkana people of northern Kenya dubbed it an 'ostrich-camel lizard,' which accurately reflects the animal's shape. Hadrosaurs had made their appearance during middle Cretaceous time in North America, Europe, and central Asia, but the more ancient forms are in general poorly known.

Hadrosaurs were bipedal walkers that bore only three blunt, hoof-tipped toes on their feet. Their necks were graceful and birdlike, and their tails were relatively long, deep, and inflexible. The teeth were grouped into elongate, inclined grinding mills. A complex system of muscle, bone, and tendon allowed the cheeks to spring apart as the lower jaws were closed, prolonging the duration of each grinding stroke. Some varieties also possessed a ducklike bill. Adult hadrosaurs ranged between 7 and 16 metres in length, and 2 and 13 metric tonnes in weight. Most were closer to the smaller limit in size.

The second most abundant group of dinosaurs in the

late Cretaceous of North America was the ceratopsians, or horned dinosaurs. In some areas they seem to have been even more common than hadrosaurs. Probably a few tens of millions of years previously, these rhino-like herbivores had evolved from bipedal ancestors, and their bodies had become compact and barrel-shaped. By dinosaurian standards their heads were huge. Their beaks were high and narrow, and a ball joint between the skull and the short neck allowed the animals to turn their beaks parallel to the ground. The extremely powerful jaws were adapted to chopping up plant materials, but they lacked the spring-grinding mechanism of hadrosaurs. The horns and bony shields extending over the neck were obviously used in self-defence, and the presence of puncture wounds in some shields indicates that the enemy could be another horned dinosaur. The broad feet bore 5 digits in front and 4 behind. The tail was relatively small and round. Adult ceratopsians ranged between 4.5 and 8.5 metres in length and 3.5 and 19 metric tonnes in weight.

A third group of plant-eating dinosaurs, which migrated to this continent from Asia probably during late Cretaceous time, was not abundant but was quite interesting. This was the ponderous, heavily armoured ankylosaurids. A bony cuirass extended from their eyelids, cheeks, and plated nostrils, across their broad back, and onto the tail, which, in contrast to that of the armoured nodosaurids, ended in a club. The bones of the pelvis and the ribs (and in at least one instance those of the shoulder girdle) were immovably fixed together to support the body shield. Impressions of soft tissues preserved on the bones of the skull indicate that the animals had poor eyesight and a keen sense of smell. Complicated, scroll-like bones in the snout possibly supported membranes that increased the humidity of inhaled air or served as heat-exchangers to cool blood flowing to the brain. A squared-off muzzle suggests that the animals, alone among their large dinosaurian contemporaries, cropped small plants growing on the ground. With a short neck and rigid back, they probably moved their heads among herbaceous vegetation by movements of their forelimbs. Low metabolic rates and a sluggish mode of existence are suggested by a very weak dentition, a relatively small brain, and a passive mode of defence. Adult ankylosaurids ranged between 6 and 8.5 metres in length and 2 and 6 metric tonnes in weight.

The carnivorous tyrannosaurids were undoubtedly a plague for plant-eating dinosaurs. These large-headed bipeds, with their dagger-filled jaws and tiny, two-fingered forelimbs, are among the most famous of all dinosaurs. However, their skeletons have been identified with certainty only in North America and central Asia. Other groups of carnivorous dinosaurs were more common in Africa, South America, and India. Tyrannosaurids are also restricted in time to the last dozen million years of the dinosaurian era.

Tyrannosaur bones evoke an image of a terrifying predator with a streamlined and efficient body structure. A brutish, bulldog neck, powerful chest, feline hips, and lanky legs imply that the animals were active hunters. An interlocking array of riblike bones protected their bellies from erratic horn thrusts of hapless herbivores. A large pelvic bone with an expanded end probably supported the animal's body in a prone position, buttressed on each side by the flexed hind limbs. Lying next to a toppled tree or among high vegetation, tyrannosaurs, like lions, could wait in ambush for their prey. There were large, wartlike growths on the ridge of the muzzle. Their function is difficult to imagine, unless they served to enhance an already stunningly repulsive appearance. Is the antithesis of beauty a matter of consensus between hadrosaurs and humans? Widely spaced teeth lining the sides of their jaws suggest that tyrannosaurids were rather messy feeders. They did bear tiger-like nipping teeth in the front of their upper jaw, which enabled them to clean a carcass somewhat more effectively than other large flesh-eating dinosaurs could do. The several varieties of tyrannosaurs ranged between 9 and 12 metres in length and 1.7 and 5 metric tonnes in weight.

The last of the five groups of typical late Cretaceous dinosaurs on this continent was the ornithomimids, or ostrich dinosaurs. The ancestry of the group is remote and extends back to late Jurassic time, when the South Atlantic Ocean had not formed and African and North American dinosaur faunas mingled. Ostrich dinosaurs in a sense can be considered tyrannosaur parodies. In details of the construction of the feet and in the presence of tiny wartlike pits on the nose they resembled tyrannosaurs. But where the tyrannosaurs had teeth, ostrich dinosaurs had none. Tyrannosaurs had large bodies and tiny forelimbs, while only the eyes and the forelimbs of ostrich dinosaurs were large. Where a tyrannosaur would attack, an ostrich dinosaur would flee. Indeed, the ornithomimids were among the fleetest of dinosaurs, evidently possessing the speed and brains of modern ostriches. Although such a proportionally great amount of cerebral equipment was quite respectable by dinosaurian standards, zoo-keepers consider ostriches the most stupid of the large animals under their care. The arms of ostrich dinosaurs were not unlike human arms, although the wrists were less flexible and the three digits of the hand were longer and ended in claws. The animals were probably omnivorous, subsisting on eggs, small animals, or fruits. Adult ostrich dinosaurs ranged between 3.5 and 4.5 metres in length and between 10 and 150 kilograms in weight.

Inauspicious Beginnings

The dinosaur remains buried under the shallow marine seaways of late Cretaceous time are generally too incomplete to reveal satisfactorily the form of the dinosaurs from which they were derived. They do document the existence of ornithomimids, large theropods, spiny-shouldered nodosaurs, and hadrosaurs on lands bordering the coasts. Two relatively complete hadrosaur skeletons, from the Kansas and Alabama chalks respectively, belong to varieties (*Claosaurus, Lophorothon*) that have not been identified in younger floodplain deposits. At a few localities fragmentary skeletal remains have been recovered from continental sediments that are approximately the same age as the chalks. In southern Alberta (Milk River Formation, 30 metres thick) these have produced the bones of ceratopsians, a small skull-cap of a dome-headed dinosaur or pachycephalosaur, and a bone from the pelvis of a large unidentified herbivore. In western Montana (lower part of the Two Medicine Formation, 100 metres thick) hadrosaur and ceratopsian skeletal parts appear to have been derived from relatively primitive members of both groups.

Dinosaur bones of the same period have also been discovered in red mudstones deposited on a floodplain near the ancient northwest coast of Mexico (Baja California del Norte, La Bocana Roja Formation). One very incomplete specimen belonged to a carnivorous dinosaur (*Labocania*) that weighed on the order of a metric tonne. The animal's limbs and head were more massive than those of tyrannosaurs of comparable body size. Peculiarities in the few preserved skeletal elements indicate that the animal was not a tyrannosaur, but its relationships to other large carnivorous dinosaurs are not clear. Bones of hadrosaurs, as well as those of a small bird named *Alexornis*, have been recovered from the same strata. The latter belonged to a group of land birds that had a shoulder joint similar to that of pterosaurs. They were good fliers but nested on the ground. These primitive birds, known as enantiornithines, inhabited North and South America, Asia, and Australia during Cretaceous time.

While fragments of dinosaur carcasses periodically settled to the floor of the sea in North America, a tropical coastline extended across what have become the deserts of Soviet central Asia. Tongues of the sea separated rank upon rank of headlands descending from hills to the east. Where the storied city of Samarkand now rises, pterosaurs then glided over bays, planing toward the surface to skim fish from warm, turquoise waters. Between the headlands, meandering streams nourished forests of broad-leaved trees lining lowlands at the heads of bays. The streams and stream banks were inhabited by fishes, salamanders, frogs, lizards, turtles, and crocodiles, all of which would have impressed a naturalist as being very similar to animals inhabiting younger dinosaur-age marshlands in North America.

The dinosaurs of Samarkand were also oddly non-exotic. There were ornithomimids and small and large carnivores that would not have seemed out of place in North America. Fragments of horns and jaws suggest that ceratopsians may also have been there. Hadrosaur remains are relatively abundant and sometimes well preserved. They belong to two distinct subgroups, which may be characterized as 'flatheads' (hadrosaurines) and 'tubeheads' (lambeosaurines). In flatheaded forms, represented by *Aralosaurus*, hollow crests are never present on the skull, the beak is relatively broad and ducklike, the spines of the backbone tend to be short, a pelvic 'foot' is absent, and the shins are relatively short. In tubeheaded forms, represented by *Jaxartosaurus*, hollow crests are present on the skull, the beak is relatively narrow, the spines of the backbone are often quite long, two slender rods from the pelvis project down and back to end in a 'foot,' and the shins were relatively long. In North America tubeheaded hadrosaurs have not yet been identified in strata of this age, although they later became abundant here. Mysterious giant claws were separated from the undiscovered skeleton of an otherwise unknown dinosaur and buried by the sands of Soviet central Asia.

Dinosaur specimens from shallow sea floors in North America and bays in central Asia are typically neither abundantly preserved nor very complete. In both cases they occur in strata deposited between 90 and 75 million years ago. North American specimens in which several skeletal elements are preserved often seem to differ from related forms more typical of later time on this content. Those from central Asia are no more different in spite of a distant origin, and skeletal parts of tubeheaded hadrosaurs are preserved there in some abundance. About 75 million years ago giant fishes apparently disappeared from the interior sea of North America, and highly evolved marine lizards simultaneously replaced their more primitive predecessors. There is thus evidence for environmental changes that were similar in magnitude and in timing both on land and in the sea. And there is mystery in these changes.

Cordilleria and the Continent

By late Cretaceous time North America had long been a continent, but it was far different from the continent we know today. The combination of a much greater extent of coastal environments and influence of maritime climates, as well as greatly reduced seasonality, would have made modern North America seem much more continental, like central Asia, according to the standards of its late Cretaceous inhabitants. Greenland was linked by land to northern Eurasia through the islands of Spitsbergen, north of Scandinavia, and the spreading axis of the North Atlantic rift system temporarily passed through Baffin Bay to the Arctic Ocean. South along the Atlantic seaboard, and the northern margin of the Gulf of Mexico, the edge of the continent was essentially the same as it is now, although the coastline passed inland from New York City through the vicinity of Washington, DC, Charleston, South Carolina, and Columbus, Georgia, and curved to the northwest through Tuscaloosa, Alabama, toward Memphis, Tennessee.

The eastern part of the continent was separated from western lands by the great interior seaway, which was more (often far more) than 500 kilometres wide, and crossed eastern Mexico, the plains states, the prairie provinces, and the Northwest Territories to enter the Arctic Ocean. At various times, arms of the interior sea isolated the Ozark Highlands and the northeastern part of the Canadian Shield from the main part of the Appalachian land mass. The biological consequences of the division of the continent into two major land masses must have been substantial.

Even the mountains were not immutable. A perusal of a topographical map of North America reveals that the

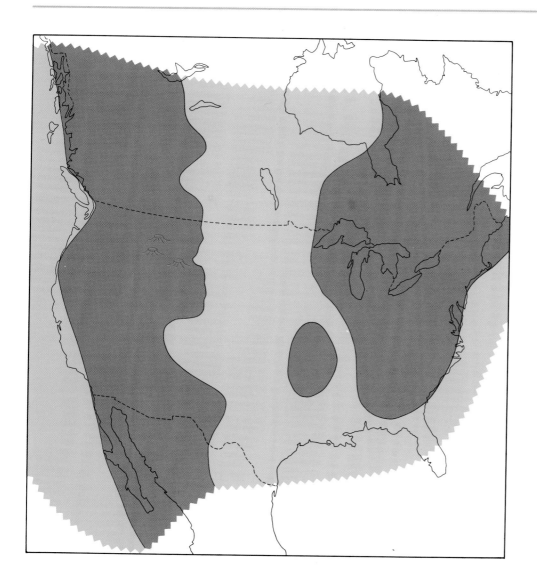

75 million years ago

North America and its interior sea. Former Pacific island terrains are being crushed to form Rocky Mountains. Along western margin of interior sea, eroding highlands produce deltas containing abundant dinosaur remains.

Canadian Rockies are unique. Nowhere else on the continent can one fly over such a vast expanse of formidable terrain, composed of range after range of spectacular mountains. They rise from highlands in northern Washington, Idaho, and Montana and continue north through British Columbia and Yukon to fan out in Alaska south of the Brooks Range. At the beginning of Cretaceous time the Canadian Rockies were not part of the continent. They were coalescing from volcanic island arcs brought together through the convergence of great crustal plates beneath the Pacific Ocean, a thousand kilometres or more to the southwest, off the coast of California.

Like New Zealand, New Caledonia, and other large islands in the southwestern Pacific today, this newly amalgamated mini-continent, which has been named Cordilleria, probably carried a distinctive assemblage of indigenous plants and animals. Many were probably living fossils of their day, and the remainder would have been carried to it by chance on natural rafts of vegetation floating on ocean currents or perhaps on the wings of giant pterosaurs. In modern times, island organisms often do not compete well with those inhabiting continental land masses. So far no record of land life has been found on Cordilleria from when it was an island.

The same kind of crustal movements that formed Cordilleria carried it toward North America, and later in Cretaceous time its northwestern edge contacted the North American land mass in northern Alberta. Cordilleria was then rotated in a counterclockwise direction, and about 75 million years ago its entire eastern coast was brought into contact with the west coast of North America. The old west coast was crumpled in three major phases over the next 20 million years, so that along the latitude of the international boundary its crust was shortened by 200 kilometres. Thus, 75 million years ago, sediments washed from old island highlands began to fill the western margin of the interior sea in Canada. To the south the sea continued to dominate the middle of North America. Perhaps the collision of Cordilleria created land-bridges to Asia that allowed tyrannosaurs, ankylosaurids, and tubeheaded hadrosaurs to migrate to this continent. And perhaps the resulting changes in atmospheric and oceanic circulation precipitated the turnover of marine faunas in the interior sea.

The Weather in Camelot

The single most important factor governing the climate of a region is its latitude. During late Cretaceous time the North Pole was located 2500 kilometres away from its present position. It was situated directly on the Alaskan-Siberian land-bridge, and the entire latitudinal grid was displaced so that the west coast of North America lay relatively to the north, and the east coast relatively to the south, of their present positions. Climatic zones were similarly shifted. Animals migrating to Asia from western North America would have had to cross the North Pole. The route from eastern North America to Eurasia was located 3200 kilometres further south. Alaska and Yukon lay entirely within the polar region. The northern interior plains were at about the same latitude as they are today, while the Appalachian region bordered on the tropical zone.

Differences in climates may be exemplified by the kinds of plants that grow under various climatic conditions. During late Cretaceous times, plants were generally only distantly related to their modern counterparts, so that climatic conditions can better be deduced not on the basis of the habitats of related living species but rather by examining the adaptations of the ancient plants themselves. Thus small vessels in wood can indicate freezing temperatures, large ones can indicate rapid evaporation in warm air, and growth rings record conditions only seasonally favourable for growth. Leaves are very sensitive indicators of air temperature, humidity, wind velocity, and seasonality. For example, large leaves with toothed margins indicate cool, wet climates; small ones with spines suggest dry conditions; smooth-margined leaves with drip tips occur abundantly in warm, humid environments; and long, slender leaves reflect windy conditions.

According to the plant adaptations, the temperature in the Appalachian region did not vary greatly from a pleasant 20 to 30 degrees Celsius. Rainfall was not heavy, but it was distributed evenly throughout the year. The trees were not tall and were widely spaced, so that they did not shade each other. They were well adapted to conserve water, so that humidity levels were low. Some of the taller broad-leaved trees bore long, narrow leaves, and scattered conifers towered over the bushy forestlands. There were few vines, and the larger trees probably had flat or gently rounded crowns as they do in open forests in low latitudes today. Similar vegetation still occurs in certain areas of Indonesia, but warm, subhumid environments in which seasons are lacking have essentially disappeared from our planet.

In the northern midlatitudes, in the region of what has become Montana and Alberta, the mean annual temperature was several degrees cooler. Rainfall may have been more abundant, but plant growth was thus subjected to seasonal changes in light and precipitation, and the forests were composed of an entirely different suite of trees. The leaves were of average size, and most were carried throughout the year. The trees tended to be taller, and some supported vines.

In Alaska and Yukon a large part of each year passed beneath the polar night, but there was no frost. The mean annual temperature remained near 10 degrees Celsius, and in many respects the humid climate resembled that of Vancouver. Growth was vigorous during the season of light, and the trees shed their very large leaves at the beginning of the season of darkness. Deciduous conifers, such as the bald cypress, also shed their fronds. Only a few conifers with small leaves carried them throughout the year.

North America would hardly have been recognizable during late Cretaceous time. A relatively slow decline in temperatures toward the poles, the north-south orientation of mountainous areas, and the separation of the continent into eastern and western land masses by the interior sea made north-south migration easier. On a planet-wide scale, mean annual temperatures were then about 10 Celsius degrees higher, an effect caused by a different distribution of lands and seas, a greater atmospheric 'greenhouse' effect caused by higher carbon dioxide levels, and perhaps other factors that have not yet been evaluated (such as a greater atmospheric mass). In the reduced effects of seasonality, and even in their absence, environmental conditions were strikingly different. No modern counterparts can be found for the warm subhumid and polar deciduous forests.

Seventy-Five-Million-Year-Old Eggs

Purplish and reddish-brown sediments are exposed on the high plains of northwestern Montana, near the city of Choteau, that long ago were deposited on a broad plain some 360 kilometres from the western edge of the interior sea (upper part of the Two Medicine Formation, 245 metres thick). They were shed from volcanic highlands in southern Montana and from rugged hills to the west that formed when vast sheets of stratified rock were pushed eastward with the impact of Cordilleria. The reddish hue of the sediments, and caliche bands in fossil soils within them, indicate that these ancient flatlands were semi-arid. The newly formed western highland had created a local rainshadow within a generally well-watered region. During the rainy season the verdant plain was dotted with ephemeral soda lakes, but during the dry season only vegetation growing near the streams remained green.

As might be expected, the semi-arid sediments contain few bones of fish and small aquatic reptiles. Dinosaur bones, however, occur in great abundance. Remains of immature animals are particularly common, and the scattered bones of babies and eggshell fragments are the most abundant of all. But only a few varieties of dinosaurs are represented in this wealth of fossil material, including a hadrosaur, a small herbivorous hypsilophodont, and a small carnivorous theropod. Bones of terrestrial monitor lizards are also common. Why did so many individuals of such a limited variety of dinosaurs seek a local semi-arid environment, virtually the antithesis of an oasis, within a vast region of vegetational abundance? The dinosaurs had come in order to reproduce, and they were seeking unusually well-aerated soils in which the eggs would not be drowned. The small carnivorous dinosaurs and monitor lizards had come to devour the eggs and young.

One shallow lake was large enough to retain its alkaline waters throughout the year, and dinosaurs nested near its shore. One variety, perhaps a small carnivore, buried 10-centimetre-long eggs in paired rows, many of which did not hatch. Algal banks and freshwater clams flourished on the tepid muds on the floor of the lake. A 4-metre-long wing of a giant pterosaur was entombed there, perhaps as a consequence of an unsuccessful high-speed dive on a juvenile dinosaur, an aerodynamic miscalculation, and a dramatic crash onto the surface of the lake. Floods swept large banks of sand into the lake, and these were transformed into islands when water levels rose. Some became 'egg islands,' for they offered a partial haven for small hypsilophodonts that returned to them season after season to nest.

The hypsilophodonts deposited about two dozen eggs within a circle about 1 metre in diameter. They did not make an excavation but simply deposited each 15-centimetre-long egg with its pointed apex in the ground in a manner so that no egg would touch its neighbour. The nest may have been covered with plant material. One clutch has been preserved in which each of the 19 eggs contains the skeleton of a tiny, embryonic hypsilophodont. When the animals hatched they immediately left the nest, for the bottom half of the eggs is never trampled. The young were precocious and fast runners. However, they must have remained in the vicinity of the nests some time after hatching, for the bones of juvenile individuals are often found near the nesting sites. So are the teeth of monitor lizards and small carnivorous dinosaurs, and so are the pupae of carrion beetles. Life was not easy for a young hypsilophodont.

Maiasaura:
detail of animal with its young,
Montana, 75 million years ago,
during late Cretaceous time

Good Mother Lizards

The dinosaur rookeries of Choteau, Montana, are remarkable also for nests of a hadrosaur called *Maiasaura* (good mother lizard), in deference to abundant evidence preserved there of the care these dinosaurs took of their young. Maiasaurs were rather small hadrosaurs, weighing about 2 metric tonnes and measuring 7 metres in length. They belonged to the flatheaded group and were relatively unspecialized creatures. Their beaks were broad and duck-like, and a small spine in the centre of their foreheads served probably as a support for fleshy structures used in trumpeting. *Maiasaura* bones have been found in two circumstances that provide insights into the rate at which the animals grew.

In one association, the skeletons of adults are found near nests, or with juveniles less than 3 metres long. In another, a great herd was overwhelmed in a heavy fall of volcanic ash, and the skeletons were later torn apart and reburied in a mud flow. This herd was composed of animals that belonged to body-length groups of about 3, 4, 5.2, and 7 metres. A simple explanation for this phenomenon is that the animals reproduced once each year and the size groups represent age groups of 1, 2, 3, or 4 (and over) years. Maiasaurs would thus have matured in 4 years and then, like modern birds and mammals, ceased to grow. Weights can be deduced from lengths, and the resulting growth rates imply that the animals doubled their weights each year until they became adult. These dinosaurs were growing more than 10 times faster than alligators, or at about two-thirds the growth rate of a mammal of similar adult body weight.

The absence of clearly defined growth rings in their bones confirms that the animals were growing at a high but uniform rate. Their tooth batteries were as sophisticated as those of modern hoofed mammals, and the area of the grinding surfaces was equally large in proportion to the hadrosaur's weight. Hadrosaurs evidently needed to chew about as much food per day as modern mammals do. High growth rates and the need for large quantities of food are indications of a high metabolic rate. Hadrosaurs were also remarkable for their very large eyes, often twice as large as those of ostriches, and for a brain volume twice that of modern reptiles, after allowing for body size effects. Maiasaurs, and by implication hadrosaurs in general, were at least incipiently warm blooded.

Much is known, or can be inferred, of the natural history of *Maiasaura*. In a semi-arid environment it would be best to lay the eggs during the latter part of the dry season, so that they would not be smothered by wet soils. The hatchlings would emerge with the return of the rains and a seasonal flush of soft, nutrient-rich plant food. The nests were built in silty earth away from the main stream channels. As preserved, they measure about 2 metres in diameter and are surrounded by a rim about 75 centimetres high. The nesting sites occur in colonies but are never closer to each other than one adult body length. About two dozen eggs were deposited in each nest. They were in the shape of lopsided ovals, about 20 centimetres long, and weighed about 1.2 kilograms. The eggs were of normal proportions for reptiles the size of *Maiasaura* adults.

The adults may have shielded the eggs from extremes of temperature with their bodies and thus 'sat' on their nests. Hatchlings were about 35 centimetres long and probably weighed a kilogram. They remained in the nests, trampling the eggshells into fragments, and wear on their tiny teeth indicates that they were fed there by their parents. Unlike the surrounding reddish sediments, the mud from which the nests were made is green, probably because of a change in the colour of iron minerals caused by vegetation the parents carried to the nests. Two out of a total of eight nests examined contained dead hatchlings. The young were evidently so tightly bonded to their nest that they chose to starve to death there, rather than to leave the nest to forage on their own. One of the nests contained the remains of 15 1-metre-long juveniles, 11 inside the nest and 4 outside but nearby. The fact that a few had at least ventured beyond the nest indicates that at this size they were nearly ready to leave it altogether. The pattern and rate of weight gain in larger animals when extended to hatchlings would imply that they remained in the nest about 60 days. The brood evidently followed its parents for many months thereafter, before leaving them at the beginning of the next breeding season.

An appraisal of the 'facts of life' as they applied to the domestic life of maiasaurs shows that parenthood must have been a challenge for them. If the hatchlings were sluggish, thereby efficiently channelling more energy into growth, then in order to attain a length of 1 metre within 60 days each would require about 1.2 kilograms of plant food per day. A brood of 20 would need 24 kilograms, and at this rate the parents comprising a breeding colony would be unable to find energy-rich fruit in sufficient quantity. The young would be fed leaves and shoots, and the economics of foraging practically necessitates that both parents fed them. It would have been easier for the parents to grind up the food and carry it to the nest within their bodies, depositing it near the hatchlings as a partially digested slurry, either by regurgitation or by elimination. A complex series of interlinking problems made foraging a dangerous and exhausting process.

The energy needs of a foraging hadrosaur can be estimated from its probable weight and from the normal walking speeds of large bipedal dinosaurs calculated from trackways. A distance of about 23 kilometres was about as far as a maiasaur could walk and still return to its nest in one day. In order to forage, to maintain its own health, and to feed its large family, each parent had to collect perhaps 75 kilograms of fodder per day. Nearby sources of food would soon be depleted, and the parents would have to range ever more widely each day. Because the nests were immobile, the parents could not disperse in order to avoid predators. One means by which prey animals minimize losses from predators is to gather in such great numbers that the proportion of animals being killed by carnivores living in the region becomes relatively small: the number of prey animals suddenly appearing in the nesting area greatly exceeds the appetites of the local population of predators. It was thus advantageous to have the greatest possible number of nesting animals in a rookery. Given the range and food requirements of the hadrosaurs in a colony, and the ability of plant growth to produce food in a semi-arid climate, an area with a radius of about 23 kilometres should have supported a maximum of about 2700 warm-blooded hadrosaurs.

The annual nesting of *Maiasaura* took on an aura of high drama. The largest breeding herd would have formed a caravan of about 2700 animals. They would have congregated in a dry alluvial plain under a burning sun and, with much dust, constructed 1350 nests over an area two-thirds of a kilometre in diameter. The rains came, the brushlands greened, and the nests burgeoned with stumbling, chirping hatchlings. Lines of parents streamed along muddy, dung-filled paths to and from the colony. The air was rent by sounds of breaking branches, as trees and shrubs were stripped of their foliage in an ever-widening circle of environmental destruction. The activity attracted the attention of every carnivore, large and small, within the region of the rookery. The fact that two out of eight nests were found to contain dead hatchlings poses the question of why the parents abandoned their nests. Were they victims of tyrannosaurs stalking the fringes of the nesting colony? Such a great toll of adult maiasaurs would quickly have to be replaced in order to sustain a breeding population and would in turn explain why they had to grow up so quickly. It may have been that a 10-year-old *Maiasaura* was a very ancient animal. Carcasses of dead hadrosaurs began to dot the stripped shrublands, and the area began to acquire something of the appearance of a battlefield.

The hatchlings grew, but so did the number of carnivores and the distance to unbrowsed vegetation. And the pulse of plant growth waned with the declining rains. The moment when the juvenile hadrosaurs were able to leave the nest probably coincided with the onset of conditions when they could no longer have survived in the nest. The *Maiasaura* parents departed with their broods toward greener pastures. Later in the year the hadrosaur families dissolved, but they retained their social habits. The bones of the great herd overwhelmed by a heavy fall of volcanic ash are spread over an area of 1¹/₂ square kilometres. In it are skeletal parts belonging to an estimated 135,000 individuals. No environment could have supported a stationary group of this many large, warm-blooded herbivores. These are the remains of a migrating herd of dinosaurs.

The plain on which the *Maiasaura* nests were made was located close to the western mountains. The plain descended to the north, and a few kilometres south of the international boundary, on the Blackfoot Reservation in Montana, another series of badlands has formed in its sediments. This site is very rich and will surely yield important information on dinosaurian ecology. Eggshells occur there in great abundance, as well as the bones of numerous other dinosaurs, but none has been identified as belonging to *Maiasaura*. Among the dinosaurs present were the hypsilophodont that nested near the *Maiasaura* site, as well as a new spiny-frilled, horned dinosaur and a small, hornless 'horned' dinosaur, or protoceratopsid. None of these is known to occur in lower, more northerly regions of the plain. However, remains of most of the other dinosaurs have also been identified in a great fossil field in Alberta.

A Park for Dinosaurs

As it crosses the southern Alberta plains, the Red Deer River passes through a vast region of badlands some 150 kilometres east of Calgary and 350 kilometres north of Great Falls, Montana. The sands and silts exposed there (upper part of the Judith River Formation, 90 metres thick) have produced more dinosaur skeletal fragments than any other comparable area of badlands on Earth, documenting the presence of twice as many dinosaurian varieties. Much of the nearly 100 square kilometres of badlands is incorporated into Dinosaur Provincial Park, which has been designated a UNESCO World Heritage Site.

Seventy-five million years ago, the badlands were part of the river system on which the *Maiasaura* nesting grounds were situated, over 300 kilometres away to the southwest. In Montana streams were deflected to the northeast by a ridge of low hills separating them from the interior sea. The hills were completely buried north of the international border, and streams turned to flow more directly toward the east. The region of Dinosaur Provincial Park was still 100 kilometres from the sea. The coastal plain was so flat, however, that the damming effects of high tides periodically slowed the flow of streams within the park. Stream channels averaged 50 metres wide and 4 metres deep. The larger channels were a third of a kilometre wide and 15 metres deep, and each complete river bend was separated from its neighbours by a distance of about 4 kilometres. These large rivers provided easy access to the sea, and elasmosaurs entered the estuaries to swim as far inland as the park.

The buried ridge of low hills probably had some influence on local environments. Streams deposited less sediment over the subterranean ridge and usually left behind sands instead of silt. Nutrients to sustain plant growth are not so readily available in sandy, well-drained soils. To both the west and east of the ridge, adjacent flatlands retained fine-grained sediments deposited by frequent floods. The silts, already enriched with clays derived from volcanic ash, provided a substrate for fertile soils. The mountains of Cordilleria lay 320 kilometres to the west, and their rainshadow effect did not extend to the park. The lush lowland plains on either side of the buried arch thus provided abundant fodder for dinosaurian herbivores.

An abundance of fossil pollen and spores is preserved in the sediments. These tiny grains were derived from plants of many different varieties, nearly half of them flowering. Members of the bald cypress family were common, as were plants allied to the modern parasitic mistletoes, cycads, tree ferns, and the often-herbaceous lilies. Other interesting plants included bushy podocarps (bonze pines), katsura trees, and members of the cashew–poison ivy family. Few leaves entered the fossil record, possibly because of soil trampling by dinosaurs. Those preserved were shed by conifers, broad-leaved trees, and vines. Broken sections of bald cypress logs measuring up to 60 centimetres in diameter and over 17 metres in length lodged on ancient sand-bars. The abundance of fossil conifer wood suggests that large trees flourished in the region. One stump has been found that belonged to a pineapple-like 'cycad' (bennettitalean) which grew so abundantly during the older part of the dinosaurian era.

Taken as a group, these plants required warm, humid climatic conditions. Bald cypresses shed their fronds during the season of short days, when plant growth in general must have slackened somewhat. Flowing streams mixed the plant debris uniformly across the region and blurred the record of plant varieties that may have preferred sandier soils over the buried ridge. Ancient environments in the park were not silent, nor was the vegetation in pristine condition. Browsing dinosaurs were a continuing source of severe trauma to the plants, and the destruction they wrought

Dinosaur Provincial Park,
north of Brooks, Alberta: badlands of Judith River Formation.
Dark strata capping badlands in distance are coaly sediments
transitional to overlying marine Bearpaw Shale.

Corythosaurus:
detail of animal in bayou,
Dinosaur Provincial Park,
75 million years ago

would have created the openings for the light that was so abundantly available to plants of all sizes. The region was probably littered with broken branches and received a rain of dung amounting to 500 metric tonnes per square kilometre per year. Footprints and trails criss-crossed flatlands that in many areas would have been strongly reminiscent of nothing lovelier than a barnyard.

More than 30 different varieties of dinosaurs have been identified in the park. The fauna, however, was dominated by only eight kinds of large dinosaurs (*Euoplocephalus, Chasmosaurus, Centrosaurus, Corythosaurus, Lambeosaurus, Kritosaurus, Prosaurolophus* and *Albertosaurus*). The remainder were either small or relatively uncommon. When mature, the large dinosaurs averaged between 2 and 3 metric tonnes in weight, but none was as heavy as a large bull elephant. Two-thirds of them were bipeds. With a respectful nod in the direction of the kangaroos of Australia, modern big-game faunas are overwhelmingly dominated by quadrupeds. Because most of the large dinosaurs were smaller than the dominant animals in brontosaur-dominated faunas, and because they were at least incipiently warm-blooded, the quantity of dinosaurs expressed as weight of living flesh per unit area was much less. There were at least 11 varieties of small dinosaurs weighing 150 kilograms or less. These were active animals that depended on their agility and often highly developed central nervous systems (by reptilian standards) for survival among more powerful contemporaries.

The dinosaurs of the park were an evolutionarily sophisticated group of dinosaurs, but because of the 'polishing' effects of millions of more years of natural selection, the mammalian big-game faunas of east Africa today are

Head of a hadrosaur rib,
Dinosaur Provincial Park

evolutionarily more sophisticated still. Although punctuated by moments of drama in carnivore-herbivore interactions, animal life in an east African game preserve moves at a relatively relaxed pace. Thus the large animals inhabiting Dinosaur Provincial Park could only have exceeded modern large mammals in the slowness and deliberation of their movements. Their world was for the most part one of languor, broken infrequently by moments of high terror and brutality.

These were members of the megafauna, the bony remains of which commonly occur as articulated skeletal fragments or mixed in dense accumulations called bonebeds. The two other categories of animal life in the park are distinct from the megafauna and each other in both size and mode of preservation. The mesofauna included animals in the size range of the small dinosaurs, and their skeletal remains are usually both incomplete and scattered. The microfauna is represented by isolated bones and teeth of relatively tiny terrestrial and aquatic vertebrates occurring typically as lag deposits at the bottom of stream channels.

More than 250 articulated specimens of large dinosaurs have been collected from park badlands, and more continue to be excavated at a rate of about six per year. Half of these belonged to hadrosaurids, over a quarter to horned dinosaurs, and the remainder to armoured dinosaurs (of both the ankylosaurid and nodosaurid kind) and to the carnivorous tyrannosaurids. The nodosaurids, with their long, slender tails and heavy shoulder spines, were of ancient derivation and were no longer as abundant as they had been. The animals (*Edmontonia, Panoplosaurus*) browsed on low plants and may have preferred flatlands in the eastern part of the park which were nearer to the sea. Heavily armoured ankylosaurids (*Euoplocephalus*) were the most common armoured dinosaur. Preferring to defend themselves actively by means of a large tail club, they probably frequented open environments where the growth of ground-dwelling plants on which they fed was luxuriant.

Ceratopsian, or horned dinosaurs belonged to one of two major kinds. The rhino-like centrosaurines (*Centrosaurus, Monoclonius, Styracosaurus*) possessed a deep, short muzzle, short neck shield, and a large nose horn well suited to inflict damage on the abdomens of tyrannosaurids. The varieties are distinguished on the basis of ornamental horns lining the rim of the neck shield. Chasmosaurines (*Eoceratops, Chasmosaurus*) bore a pair of relatively long horns over the eyes, but the horn on the nose was short. The neck shield was very long, and its appearance was enhanced when the animal lowered its head to direct the brow horns forward. Chasmosaur skulls have been found bearing partly rehealed horn wounds, showing that the animals fought with each other. It is speculated that the brow horns of females sloped forward over the muzzle while those of males were more upright, causing the frill to be more prominently displayed when the horns were lowered. These animals had powerfully constructed elbows, suggesting that they used their forelimbs to position and thrust the head. Few dinosaur bones have been recovered from sediments deposited on wetlands between rivers, but those that have been found usually belong to ceratopsians. Perhaps the animals preferred these areas to the margins of streams. Associated skeletal fragments of subadult ceratopsians are rarely found, one spectacular exception being a skull of a half-grown *Monoclonius*.

The most abundant large dinosaurs in the park were the tubeheaded lambeosaurine hadrosaurs (*Corythosaurus, Lambeosaurus, Parasaurolophus*), articulated skeletal fragments of which outnumber those of the flatheaded hadrosaurines by nearly two to one. The most characteristic part of the lambeosaurine skeleton is the hollow crest ('coxcomb') on top of the skull. If these were sound-producing organs, as seems likely, then trumpeting must have enabled widely separated individuals of a particular variety to identify each other. The hollow crests emitted low-frequency sounds that would have carried well over great distances. In *Corythosaurus* 'males' the rounded crests are larger and more inflated than those of 'females,' while *Lambeosaurus* 'males' bear a backwardly projecting rod that is absent in 'females.' The crests were not formed in hatchlings and began to develop when the juveniles attained a length of between one-third and one-half that of an adult. They thus very likely aided males and females in recognizing each other, the signal being acoustic in corythosaurs and visual in lambeosaurs. Only one specimen of the spectacular 'trombone-crested' *Parasaurolophus* has so far been identified in the park, and the occurrence must be accidental.

Remains of flatheaded hadrosaurines (including the rare *Brachylophosaurus* and more common *Kritosaurus* and *Prosaurolophus*) are only one-third as abundant as those of lambeosaurs west of the buried ridge of low hills but occur in equal numbers in sediments deposited each of its crest. Perhaps they preferred environments nearer to the sea, rather than the peculiarities of more local conditions on the sandier soils over the crest. Bones of the related *Maiasaura*, if they occur at all, are surely uncommon.

Although they were living at the same time as *Maiasaura*, the park's hadrosaurs were geographically and probably ecologically separated from their inland cousins. There are indications of similarities in life cycle, however. Bones of very young hadrosaurs and bits of eggshell do occur in the park, but they are very rare. The park hadrosaurs thus nested elsewhere. Although the evidence is incomplete, it would appear that individuals of *Lambeosaurus* can be separated into four size-age groups, as in the *Maiasaura* herd. Perhaps the animals were non-residents and migrated through the park once each year, losing individuals when they were in the same four stages in their growth. In *Corythosaurus* the size-age groups appear less distinct, and the animals may have spent sufficient time in the park each year to qualify as 'residents.' In view of the reality of dinosaur migration, the large number of different kinds of dinosaurs preserved in the park may in part be caused by proximity to ancient migration routes. Perhaps the tendency of hadrosaur skeletons to occur in banks of stream-transported sands reflects the tendency of migrating animals to follow more open stream-side routes.

One group of large dinosaurs is conspicuous by its absence. Although sauropods have been found in sediments of this age deep in the heart of central Asia, they are not known to have reached latitudes as far north as that of southern Alberta during late Cretaceous time. Their remains have not been identified in Dinosaur Provincial Park. In Jurassic faunas in North America, the sauropods were capable of reaching high into trees to feed. Most of the park's herbivores browsed in bushes and trees at heights ranging between 1 and 4 metres off the ground. This was very different from the situation in east Africa today, where most large mammals graze on low herbaceous vegetation. Perhaps plant-animal interactions over long periods of geological time have focused the feeding of large herbivores at ever lower levels.

A poached rhino kill,
near Lake Manyara,
Tanzania

There were two varieties of large, 9-metre-long tyrannosaurids in the park. Remains of the slender *Albertosaurus* occur in the same proportion to those of the more powerful *Daspletosaurus* as do those of hadrosaurs and ceratopsians. Perhaps these ratios reflect the prey preferences of the carnivores; possibly the tyrannosaurs had the same high metabolic rates as hadrosaurs. However, as many as 8 per cent of the skeletal fragments collected pertain to tyrannosaurs. The resulting predator-prey ratio would be impossible to maintain if the carnivores were warm-blooded, because there would not be enough prey animals available to fuel their energy-demanding metabolisms. Either the tyrannosaurids had lower metabolic rates or their remains are over-represented in the park.

Disarticulated bones of large dinosaurs often occur jumbled together along certain levels in fossil sandbanks. Over one hundred bonebeds have been identified in the park, and many of them were caused by meandering streams cutting into their banks and disrupting skeletons buried there. The bones were then washed away and tended to accumulate in those portions of the stream bed where currents slackened. The abundances of the different varieties of dinosaurs represented in them thus correspond to their relative abundances in skeletal fragments. A few bonebeds, however, are strikingly different. They are dominated by the bones of one kind of horned dinosaur, such as *Centrosaurus*, *Monoclonius*, or *Styracosaurus*.

One of the largest bonebeds is spread across a third of a

Centrosaurus
nose horn and frill,
Dinosaur Provincial Park

hectare and contains disarticulated bones belonging to an estimated minimum of 333 *Centrosaurus* specimens. The following series of events is postulated to have taken place. A *Centrosaurus* herd attempted to swim across a large stream but, because of panic, current speed, turbulence, or for some other reason, a large number of animals drowned. Hundreds of carcasses were stranded along the banks of the river, where they putrefied. Many hundreds of tonnes of decaying flesh attracted scavenging tyrannosaurs. They trampled the skeletons, breaking bones and cutting grooves into them with their teeth. Some of their own teeth broke off and were left behind, scattered among the ceratopsian bones. Then the banks were undercut, and some bones were tumbled into the centre of the stream channel where they were fossilized.

Animals belonging to three size-age groups are preserved among the specimens derived from the *Centrosaurus* herd, implying that they grew to maturity within three years. This is evidence of very rapid growth rates. Again, as in the case of hadrosaurs, the relation between the area of the dental batteries and body weight is also similar to that in mammals. Ceratopsians too were at least incipiently warm-blooded. The ratio of subadults to adults implies that 12 per cent of the adults died each year and that an average lifespan of an adult was 8 years. If the numbers of animals represented in the infant and subadult classes are changed to compensate for the loss of small bones, the number of animals in each category becomes equal, which would appear to be biologically reasonable. However, 20 per cent of the adults would then have been replaced each year, and the average life-span of an adult *Centrosaurus* would be reduced to 6 years.

Surprising as it may seem, hadrosaurs and ceratopsians could have completed their life cycles in less than a decade. In a zoo, a modern mammal of similar body weight can be expected to live for 40 years, and it is common knowledge that large farm animals usually live much longer than 10 years. However, these dinosaurs were not living in a zoo. They were exposed to the hazards of a free life and heavy predation from tyrannosaurs. Studies of reproduction and growth in modern large mammals indicate that, in the wild, mammals weighing 2 metric tonnes would replace more than 10 per cent of their population each year. Like modern mammals in the wild, the better analogue for a late Cretaceous hadrosaur or ceratopsian would be a heifer raised for beef, not the family dairy cow.

Animals that weighed between a few and over 100 kilograms may be considered to have belonged to the mesofauna, and in decreasing order of abundance these were turtles, small dinosaurs, crocodiles, and gavial-like reptiles called champsosaurs. Many of the turtles belonged to the baenids, an extinct group, which were aquatic in their habits and unusual in that they possessed very long tails. Soft-shelled turtles (trionychids) were very common. One large variety (the dermatemydid *Basilemys*) was a terrestrial, tortoise-like form that grew to be about a metre long. The crocodilians, which preyed on aquatic turtles, and the champsosaurs, which fed on fish, were not large, averaging less than 2 metres in length. Both the turtles and crocodile-like reptiles are much more commonly found to the west of the buried ridge, where quiet-water sediments are also preserved in greater abundance.

Associated with the mesofauna are the water-worn cranial domes of small herbivorous dinosaurs called pachycephalosaurs (*Stegoceras*, and very rarely *Gravitholus*, *Pachycephalosaurus*, and *Ornatotholus*). These peculiar bipeds were generally small and had flattened, discoidal bodies and tails that were protected by a latticework of subcutaneous bones derived from tendons. They bore a large dome on the top of the skull that was often bordered by spines. The arrangement of spines differed by variety. The domes of 'male' individuals of *Stegoceras* are more massive than those of 'females,' and both occur in equal abundance. If one were counting just cranial domes, *Stegoceras* would be the most common dinosaur in the park. These solid structures, with which they probably butted each other, were seldom eaten and very resistant to abrasion. The true relative abundance of *Stegoceras* is indicated probably by a single fragmentary skeleton, which, like the numerous isolated domes,

was found west of the buried ridge. Although other small plant-eating dinosaurs, such as hypsilophodonts, thescelosaurids, and protoceratopsids, lived in the lands west of the interior sea, their remains have not been identified in park sediments. They were evidently excluded for ecological reasons.

The small carnivorous dinosaurs were associated with their larger relatives, and their distributions were not obviously affected by the buried ridge. Taken as a group, probably they were either scavengers or fed frequently on creatures that were in turn consuming carrion left behind by the tyrannosaurs. Numerically, about one-quarter of all dinosaur skeletal fragments in the park belonged to small adult dinosaurs, and this fraction was probably closer to one-third in life. Nearly three-quarters of the small dinosaurs were the fleet ostrich dinosaurs, or ornithomimids (*Struthiomimus*, rarely *Dromiceiomimus* and *Ornithomimus*). The animals lacked teeth in their birdlike bills, and *Struthiomimus* possessed particularly strong forearms ending in molelike claws. They seem to have been well adapted to feed on soil- or dung-dwelling insects, salamanders, and other small vertebrates, which to judge from the numbers and size of ornithomimids must have been quite extensively available. Dinosaur eggs were uncommon, and it seems unlikely that they were a prime source of ornithomimid nourishment, although turtle eggs may have been.

The most abundant small 'Cretaceous coyote' was an interesting small dinosaur called *Troodon* (formerly *Stenonychosaurus*). It possessed many distinctive attributes, including a talon on the inside of its foot, a tail that it used as a baton to control high-speed turning movements, raptorial, three-fingered hands that contained an opposable digit, and enormous eyes. Most significantly, the brain cavity was enlarged. The brain was much larger than those of living reptiles of comparable body weight and had attained the proportions of some living mammals and birds. The animal was an energetic and highly evolved creature and, like the coyote, was probably an opportunistic carnivore. It evidently had a taste for baby hadrosaurs, for *Troodon* teeth are frequently found associated with *Maiasaura* nests in Montana. The animal had all the equipment necessary to feed on contemporaneous rodent-sized mammals, including large eyes with overlapping visual fields that must have been useful for twilight hunting. A rare and somewhat smaller carnivore with an equally large brain was *Saurornitholestes*, the skeleton of which is very incompletely known.

Dromaeosaurus resembled *Troodon* in its general body form but had a more formidable set of claws, smaller eyes with only narrowly overlapping visual fields, and a smaller brain.

Its shape is reminiscent of what a bipedal monitor lizard might look like, and the animal fed probably on carrion left behind on tyrannosaur kills. A final small theropod was the odd *Chirostenotes*, which had a short tail, very long legs, and a broad bill lacking teeth. The animal had a very large pelvic canal and may have given birth to living young. In general form it resembled the dinosaurian equivalent of a wading bird. The skeleton of a small, closely related dinosaur was found in fossil wind-blown sands near an ancient lake in central Asia, beside a nest containing eggs laid by a small horned dinosaur.

The remaining middle-weight park reptiles were at home in the skies. A very few fragmentary and delicately constructed bones indicate that several different kinds of pterosaurs soared in rising columns of warm air over the flatlands below. One form had a wing-span of 1 metre, and another of 3.5 metres. An arm-bone as large as that of a large herbivorous dinosaur was found, but its wall is as thin as an eggshell. It belonged to a gigantic pterosaur with a wing-span estimated to have been 13 metres.

Teeth, bony scales, and small, dense bones resist the processes of digestion, decay, and erosion. They are the ultimate garbage of park ecosystems, wafted by currents along channel bottoms and collecting in shallow depressions. Although the freshwater streams were the home of fishes and aquatic reptiles, the remains of these animals are overshadowed by material washed in from the surrounding flatlands. Only a fifth of stream-bed debris was derived from fishes. The most common fishes were a rayfish (*Myledaphus*) and a gar (*Atractosteus*), followed by bowfins (*Amia*) and sturgeons (*Acipenser*). Shark teeth were rare, although they are abundant in estuarine environments elsewhere. As a whole the fishes are remarkable for having survived to the present and in that very few different kinds have been identified there, in contrast to the great variety inhabiting modern stream environments.

Two-fifths of the debris belonged to the smaller reptiles, especially turtles. Perhaps because they were the most aquatic of all, bones derived from soft-shelled turtles are abundant. Crocodile and champsosaur bones are also common. Salamander remains pertaining to three varieties occur in low numbers, along with isolated frog bones. Evidence of small land-dwelling vertebrates is rare. Monitor-lizards are sometimes represented, but no remains of snakes have so far been identified. The tiny teeth of mammals are seldom found, although they have been intensely sought and are relatively easy to identify. Possibly moist areas near heavy undergrowth were more congenial to mammals or to the preservation of their remains. Most of the teeth recovered

belonged to several different kinds of ancestral opossums, one of which grew to be two-thirds the size of the modern Virginia opossum. Some teeth identify the presence of marmot-sized, seed-eating multituberculates. These extinct mammals lived probably in trees and bore their young in pouches, as opossums do. Teeth of several rare and tiny varieties are indicative of an insect diet. The animals were related to mammals that give birth to well-developed young and are thus ancestors of most modern mammals.

Dinosaur debris is very common in the channel sands. Much of it consists of the easily shed teeth of small theropods, tyrannosaurids, and hadrosaurs. Sites enriched in fish bones contain few dinosaur teeth. Nevertheless, the flow of organic nutrients from the land into the streams surely helped to sustain aquatic communities. There is no evidence of a change in the composition of stream-channel scrap across the buried ridge. Indeed, the relative abundance of the various components changes very little in sediments of this age located far beyond the confines of the park.

Stream channel deposits, Dinosaur Provincial Park

The Old Southwest

Two hundred kilometres from Albuquerque, along the edge of the Navajo Reservation in northwestern New Mexico, a vast interior basin is now drained by the San Juan River. Here, just as the sea began to spread over the fertile coastal plain 1600 kilometres to the north in central Alberta, sediments containing the bones of dinosaurs began to accumulate along the southwestern edge of the interior sea. It

Sands deposited in
ancient stream courses,
eroding in badlands of
Dinosaur Provincial Park

An Odyssey in Time

San Juan Basin, New Mexico,
near Bisti Trading Post, Navajo Reservation.
Eroding channel sands (BOTTOM: a hoodoo),
low in Fruitland Formation;
a *Pentaceratops* skull was excavated nearby.

Near Bisti: field of small hoodoos
in Fruitland Sandstone

was hotter in New Mexico, and rainfall was less abundant. There were no seasons as we know them, and growth rings did not form in trees. A sheet of sediments (Fruitland Formation, 170 metres thick) spread from volcanic highlands to the southwest to merge with the sea in a belt of swamps 10 to 25 kilometres wide. Oyster banks grew in sandy estuaries penetrating the swamps, and the brackish waters were frequented by sharks. Between the estuaries were vast bogs, where the unstable soils could support only occasional ragged clumps of conifers. Turtles, which were closely related to northern forms, flourished in the hot, wet scrublands, but few dinosaurs ever entered them.

There were areas inland, near the swamps, where the flatlands were better drained, and their sediments contain fields of palm and bald cypress stumps. Subtropical conifers related to *Podocarpus, Agathis,* and *Araucaria* grew there, as well as several different kinds of palms and numerous varieties of flowering trees. Gently flowing streams were covered with water lettuce (*Pistia*) and inhabited by sawfish, paddlefish, and gars. Among the small backboned animals, salamanders, lizards, and mammals seem to have been present in greater diversity than in Alberta. Remains of a

small snake have also been identified. The dinosaurs are represented for the most part by isolated teeth and bones. These indicate that, as in Alberta, the fauna was dominated by hadrosaurs. Ceratopsians ranked behind them in abundance. A superb specimen of the trombone-crested *Parasaurolophus* has been collected, in which the crest is shorter and more recurved than in its Canadian cousins. The elongated form of the chasmosaurine neck shield is evident in a well-preserved *Pentaceratops* skull. A life reconstruction of this 5-tonne horned dinosaur stands on the grounds of the New Mexico Museum of Natural History in Albuquerque.

As the coastline of the interior sea retreated to the east, the coastal swamps of northwestern New Mexico were buried by advancing river floodplains (Kirtland Formation, 500 metres thick). These sediments contain less organic material than those beneath them and supported drier soils. Plant cover was more tenuous, and the remains of small backboned animals are rarely found. Dinosaur skeletal fragments occur in slightly greater abundance, however, and in greater variety. Ostrich dinosaurs and scavenging dromaeosaurs were present. Specimens of the tyrannosaur *Albertosaurus* have been recovered, as well as of the flat-

Ojo Alamo Wash, San Juan Basin. Yellowish sandstones (foreground) are part of Farmington Sandstone Member, Kirkland Formation; greyish Upper Shale Member is overlain by sandstones of Naashoibito Formation.

headed hadrosaur *Kritosaurus* and trombone-crested *Parasaurolophus*. Articulated skeletal fragments of *Pentaceratops* are common, and its smaller relative *Chasmosaurus* was probably also present. A few scattered elements indicate that the long-tailed armoured nodosaurids were there as well.

No essential difference has been identified between the dinosaurs inhabiting the back areas of the coastal swamp and those dwelling on the floodplain replacing them. The same community seems to be represented in both cases, although dinosaur remains are less common in the lower part of the sequence of floodplain sediments. The upper part of this sequence was derived from rising highlands to the north, and plant microfossils preserved within it indicate a shift toward conifer dominance and appearance of different kinds of broad-leaved plants. Whether or not there was a concomitant change in dinosaurian faunas is not known.

Many of the dinosaur specimens found in the basin of the San Juan River had ranges that extended north into Canada. However, dinosaur remains are not found here in the same abundance and completeness as in Montana and Alberta. About as many specimens have been found in New Mexico over many years as have been recovered from the Blackfoot Reservation in a few seasons. In comparing the dinosaurs of these two areas, it would appear that the Montana site contains greater variety. No rhino-horned centrosaurines have been found in New Mexico, and the armoured ankylosaurids with their tail clubs have not certainly been identified here either. On the basis of articulated skeletal

fragments, ceratopsians outnumbered hadrosaurs on the New Mexican floodplain, while the reverse is true to the north in a similar topographical environment. Quite possibly the plants of the old San Juan basin provided better fodder for horned dinosaurs, and the hotter non-seasonal climates did not provide such a benign home for dinosaurs as existed further north.

Further south still in Trans-Pecos Texas the scorched and desolate Chisos Mountains lie cradled in the Big Bend of the Rio Grande. Here, an Indian legend says that when the Great Creator had made the earth and heavens he dumped the rubble left over into what has become Big Bend National Park. South of the park, across a narrow, tepid river, spread the Mexican states of Chihuahua and Coahuila. In Trans-Pecos Texas, some 75 million years ago, lagoons and marshes also lined the margins of a series of small deltas into the southwestern edge of the interior sea (Aguja Formation, 250 metres thick). In these sediments occur the most southerly dinosaurs in the United States.

Isolated bones indicate that ostrich dinosaurs, and toothless egg-eaters allied to *Chirostenotes*, inhabited lowlands behind the seaward edge of the deltas. Large carnivores are represented by teeth. Among the small herbivores were the fleet hypsilophodonts and small, dome-headed pachycephalosaurs, but remains of these animals are rare. A long-tailed armoured nodosaurid (*Panoplosaurus*) was not uncommon. Hadrosaurs were the most abundant of the large herbivores, and their remains are represented by a range from juveniles 3 metres long to adult 10 metres in length. *Kritosaurus* seems to have been the dominant hadrosaur, but remains of a tubecrested lambeosaur have also been identified. The skeletons of about 14 individuals of the horned dinosaur *Chasmosaurus* were found at a single site. The carcasses had lodged within a channel and were derived from animals ranging from juveniles weighing about 200 kilograms to adults weighing nearly 2 metric tonnes. It would appear that a *Chasmosaurus* herd foundered in flood waters. Juvenile animals were unusually vulnerable, for their remains are nearly four times as abundant as those of adults.

The Big Bend dinosaurs closely resembled their contemporaries in New Mexico, although *Pentaceratops* has not yet been identified here. Only a relatively small amount of skeletal material has been collected, but it would appear that the Big Bend was inhabited probably by about the same number of dinosaurian varieties. In neither region were the faunas as varied as those of the northern plains. The largest animal in western Texas was not a dinosaur, but the gigantic crocodile (*Deinosuchus*), which attained lengths estimated to be of the order of 15 metres and weights

Big Bend National Park, Texas. TOP: Sunset.
BOTTOM: Exposures of Aguja Formation,
Tornillo Creek. Remains of a gigantic crocodile
(*Deinosuchus*) were excavated nearby.

well in excess of 10 metric tonnes. The huge crocodiles were quite capable of feeding on any of the herbivorous dinosaurs and, if the opportunity presented itself, even on tyrannosaurids. Although they seem to have preferred tropical to subtropical waters, fragments of their skeletons have been found as far north as Montana and are common along the Atlantic seaboard.

Scattered remains of ornithomimids, tyrannosaurs, hadrosaurs, and ceratopsians occur in river floodplain sediments of approximately this age near Saltillo, in the state of Coahuila, in Chihuahua, and in southern Arizona. Available information suggests that the dinosaur fauna of the southern plains extended with relatively little change into northern Mexico. However, these localities were located in warm climates on the edge of a tropical sea dotted with reefs and shallow limestone banks. They have not been well explored, and because of the geographical setting was not identical to that of the northern plains, it might be expected that the faunas of the two regions would not be identical.

Tropical Dinosaurs

As with Cordilleria, the entire western margin of North America was in the process of slowly acquiring island terrains from the Pacific Ocean. Its position paralleled a great subduction zone, where plates of heavy oceanic crust were descending into deeper levels of the planet. Then, as now, lands bordering the Pacific were also being torn apart by enormous zones of horizontal displacement, like the modern San Andreas fault system in California. The peninsula of Baja California in northwestern Mexico was then located 1800 kilometres to the south, in the general latitude of present-day Guatemala. Dinosaurs, the bones of which are found near the village of El Rosario 300 kilometres south of San Diego, were living well within the late Cretaceous tropics.

The badlands of El Rosario are not extensive. They occupy a basin that now parallels the Pacific coast for 23 kilometres and extends 15 kilometres inland. The sediments entering the basin (El Gallo Formation, 150 metres thick) were carried in torrents that descended from rugged coastal mountains on its eastern edge. Streams flowed rapidly through gravel bars and across a narrow floodplain to enter the ocean. Water-tumbled petrified logs 5 metres long are not uncommon, nor are stumps of uprooted trees. Palms, conifers, and broad-leaved trees flourished in the wet soils and were festooned with vines. Nearer the ancient coastline, water-worn pieces of bone become more common. Along the shore, where stream banks merged with beaches, well-preserved bones occur in the greyish-yellow silts and sands. Shells of marine creatures are found there, washed by the surf into the skeletons of dinosaurs.

Bones of small animals found in the silts include those of frogs, lizards, turtles, and birds. Associated with them in dark grey silts are the teeth of tiny mammals, very similar to mammalian teeth found in sediments of the same age in the northern plains. The similarity suggests that the animals inhabiting various regions in the western half of the continent were not ecologically isolated from each other and that the climatic regimes in mid- and low latitudes differed less from each other than they now do.

A few of the bones of larger animals belonged to tyrannosaurs and armoured dinosaurs. Most of the dinosaurian material, however, consists of bones and skeletal fragments derived from tubecrested hadrosaurs. The animals obviously figured heavily in the food chain, for out of 10 specimens examined 2 had broken-off tyrannosaur teeth associated with them, and blowfly cases were found with one. The most common form was a variety of *Lambeosaurus*, which had unusually long vertebral spines and a very high, narrow tail. The animal may have used its tail and the finlike structure extending the length of its back to radiate or assimilate heat. Because seasonal changes in temperature would probably have been minimal, perhaps the daily temperature range was relatively great. Most of the lambeosaurs were from animals averaging about 9 metres in length. However, one arm bone belonged to a giant specimen that must have been on the order of 15 metres long and weighed about 13 metric tonnes.

Hadrosaur Giants

The magnificent escarpment of the Book Cliffs extends from western Colorado far into central Utah. Within it are preserved sediments that trace the movements of the interior sea, as its shoreline withdrew to the east across this region (Mesaverde Formation, 675 metres thick) about 75 million years ago. Swamps and floodplains pursued the retreating sea but often foundered beneath it during brief readvances, when they were covered by marine clays. Dinosaurs walked across the soft peaty bogs near the shore, and their footprints were filled with silts and sands carried to them by flooding rivers. The peat subsequently changed to coal, and the sands and silts to stone. Throughout eastern Utah and western Colorado, sandstone footprint casts are preserved on the roofs of mine shafts, their contours projecting into the subterranean gloom after the coal is removed.

The footprint casts can be so dense that they merge into

a trampled field. They were made by many different kinds of dinosaurs, nearly all of which were moving with their tails clear of the ground. Tail drag marks are extremely rare. A few tracks were left behind by tyrannosaurs and ceratopsians, but most were made by bipedal herbivores. Many of these bipeds were very small animals, but whether or not they were infant or immature hadrosaurs is not known. At one site hadrosaur footprints were found among fossilized leaf litter, fallen branches, and upright, coalified tree stumps. Some animals had actually stepped on the major roots, and their footprints are positioned around the stumps in a manner suggesting that the hadrosaurs were browsing within the lower branches of the tree. A few of the hadrosaur tracks were 1 metre wide, indicating animals that weighed about 12 metric tonnes.

Virtually all the hadrosaur skeletal materials found in North America were derived from animals weighing between 2 and 4 metric tonnes. A jaw-bone collected in Wyoming belonged to an animal that probably weighed on the order of 8 metric tonnes. But the arm bone from Baja California and a few footprints from Colorado and Utah are the only available evidence of truly giant hadrosaurs on this continent. The remains of giant hadrosaurs are relatively more common in late Cretaceous sediments in central Asia. In Mongolia, a tubeheaded hadrosaur (*Barsboldia*) and a flat-headed hadrosaur with a cranial spine (*Saurolophus*) both attained lengths of approximately 12 metres and weights of 10 metric tonnes. The largest known hadrosaur was the flatheaded *Shantungosaurus* from eastern China, a mounted skeleton of which is nearly 15 metres long. The animal weighed about 12.5 metric tonnes when alive.

In living animals many ecological and behavioural characteristics are closely correlated with adult weight. These relationships can be used to pose questions on how the natural histories of 2-metric-tonne and 10-metric-tonne hadrosaurs might have differed, and the inferences thereby gained can be kept in mind when reviewing the fossil record. For example, according to the relationships, the number of large hadrosaurs in a region would be less than half that of their smaller cousins, but the weight of all the large hadrosaurs combined would be nearly twice as much. The eggs of the large hadrosaurs would be three times heavier, the number laid would be only three-quarters that of the smaller hadrosaurs, and about half as many nests would occur in each rookery. If they migrated without feeding, large hadrosaurs would consume their total bodily fat reserves at about half the rate of small hadrosaurs. Small hadrosaurs could have easily migrated over distances of one to two thousand kilometres; the large ones may have covered distances of continental proportions. If the large Asian hadrosaurs reached the Bering Straits during the arctic summer, might the few North American occurrences of the giants represent occasional visitors from the east? The correlates of body size provide invaluable tools with which to extend the reality of fossil bones and search for silhouettes of once-living creatures.

Parras Basin (west of Saltillo, Coahuila, Mexico).
Inclined strata of Difunta Group in distance;
dinosaur bones were recovered from shallow exposures
of Cerro del Pueblo Formation in foreground.

8

Triceratops

LATE CRETACEOUS,
ABOUT 67 MILLION YEARS AGO

Two great horned dinosaurs,
Triceratops, beside a small lake,
in a forest of bald cypress and
gum trees, south of Wood Mountain,
Saskatchewan.

The 'Modern' Cretaceous

A Change in Geography

Throughout most of late Cretaceous time, a well-drained lowland corridor, between the edge of the interior sea and the western mountains, extended from central Mexico to the Arctic Ocean. The corridor was over 5000 kilometres long and averaged 350 kilometres in width. It is hard to imagine a topography more conducive to the north-south migration of large animals. About 70 million years ago the continuity of the corridor was disrupted. First, a large bird-foot-delta spread to the east from eastern Idaho across northern Wyoming, and a similarly large delta invaded the eastern margin of the sea in Alberta. Then, accompanied by earthquakes, rugged hills rose from the western margin of the interior sea in central New Mexico, Colorado, and Wyoming, and smaller domes emerged in southern Alberta and Saskatchewan.

The sea retreated at least 900 kilometres to the east, leaving behind poorly drained alluvial plains between the 'island' highlands. The eastern and western shores of the interior sea may have been linked in an isthmus through South Dakota. Far to the north, alluvial fans from the Mackenzie Mountains may also have bridged the interior sea, cutting it off from the Arctic Ocean. Centred on the 49th parallel, a large remnant of the sea may have reached the North Atlantic via Hudson Bay. In addition to the profound topographic changes, an equally important climatic warming occurred at this time. Mean annual temperatures apparently rose by about 3 degrees Celsius at midlatitudes in North America. Environmental changes of this order could be expected to produce changes in dinosaur faunas.

Formerly, the sea had readvanced over southern Alberta, inundating the strata now exposed in Dinosaur Provincial Park. As it began to withdraw again, environmental changes were not initially major. Deltaic sediments (Horseshoe Canyon Formation, 280 metres thick), which were deposited in the edge of the sea in the region of the Red Deer River valley near Drumheller, Alberta, have produced numerous skeletal fragments of dinosaurs. Fewer specimens have been collected, but it would appear that about as many different kinds of large dinosaurs inhabited the delta as were present in Dinosaur Provincial Park. However, an additional habitat is represented here that was not present in the park.

These were the bald cypress swamps and peatbogs lining the seaward edge of the growing delta. The back regions of these wetlands were the preferred habitat for two large dinosaurs. By far the more abundant was a flatheaded hadrosaur (*Edmontosaurus*) that could attain a length of 10 metres and a weight of nearly 4 metric tonnes. Possibly as many as three-quarters of dinosaur specimens collected from badlands near Drumheller pertain to this form. The other peculiar animal was a large centrosaurine horned dinosaur (*Pachyrhinosaurus*) that bore a huge, crater-like boss covering most of the top of its muzzle. It approached 8 metric tonnes in weight and was thus twice as heavy as the largest horned dinosaurs from older sediments in Dinosaur Provincial Park. Abundant pachyrhinosaur remains have been recovered from a bonebed near Lethbridge in southern Alberta, associated with those of edmontosaurs. Pachyrhinosaur bonebeds have also recently been discovered near Grande Prairie in north-central Alberta. The ubiquitous terrorizers of the late Cretaceous, tyrannosaurids pursued their prey even within the bald cypress swamps.

Dinosaurs inhabiting the better-drained floodplains inland from the swamps generally resembled those of the older park. One obvious faunal difference, however, lay in the rareness or absence of nose-horned centrosaurines. Here

Horseshoe Canyon Formation, Red Deer River, north of Drumheller, Alberta.
TOP: Coal-bearing strata (low foreground), source of many specimens of *Edmontosaurus*.
Non-coaly strata (high background) show dinosaur fauna similar to Dinosaur Provincial Park.
LEFT: Coal-bearing strata.   RIGHT: From bottom: fluvial strata of upper Horseshoe Canyon
Formation; black shales of Battle Formation, deposited in shallow lake basins; well-stratified,
yellowish, fluvial strata of lower Scollard Formation, source of *Triceratops* skulls;
dark coal bounded by sandstones (upper left), with Cretaceous-Tertiary boundary

Bald cypress,
Okefenokee Swamp,
Georgia

Suwannee River,
Georgia

Hell Creek Formation,
Hell Creek State Park,
north of Jordan, Montana

the horned dinosaurs (*Anchiceratops, Arrhinoceratops*) all bore long neck shields and well-developed brow horns typical of chasmosaurines. Another equally interesting difference was probably the result of time, for evolutionary changes are evident in the skeletons of some of the dinosaurs. The lower leg is relatively longer in the tyrannosaur *Albertosaurus*, suggesting that the carnivores were becoming more fleet. The nubbin on the forehead of the flatheaded hadrosaur *Prosaurolophus* had been transformed into a graceful, unicorn-like spine in the descendant *Saurolophus*. And the hollow narial tubes on the head of the lambeosaur *Hypacrosaurus* were more completely sheathed in bone, and the crest extending along its back and tail was higher than in the ancestral *Corythosaurus*. The dinosaurian era was so long, and its fossil record so incomplete, that it is rare to find good examples of evolutionary changes in dinosaurs over a span of only a few million years.

To the south, a fauna dominated by large horned dinosaurs was already becoming established on lowlands behind a rapidly retreating sea. Soon afterward, in southern Alberta and Saskatchewan, the character of the landscape also changed rapidly, as shallow lakes spread across a basin more than 600 kilometres wide. For about 500,000 years the basin remained poorly drained and received little sediment other than wind-blown volcanic ash. Similar basins formed in Utah and Wyoming to the south. Then the lakes were filled by silts and sands belonging to an enormous alluvial plain. Although it was occasionally interrupted by marshes and rugged hills, the plain stretched for hundreds upon hundreds of kilometres in every direction. The expansion of a rich, midlatitude environment and warmer climatic conditions produced important changes in the regional dinosaurian community.

Triceratops on Montana Flats

The Missouri River has been dammed in eastern Montana, and a huge artificial lake created behind the village of Fort Peck. Sediments deposited near the centre of the ancient flatlands (Hell Creek Formation, 140 metres thick) are now exposed beautifully within Hell Creek State Park, on the southern margin of the lake. The badlands continue far to the east, around the southeastern shore of the reservoir. These strata, originally famous for having yielded the scattered skeleton of a giant carnivorous dinosaur that came to be called *Tyrannosaurus*, have recently been surveyed for remains of plants and animals that were contemporaries of the giant carnivore.

Fossil soils covering the floodplain were well drained and of a kind that form under relatively dry, subtropical conditions. Abundant traces of tree roots are preserved in them, and forests were generally dense. They were composed of many different kinds of trees. Most reproduced by means of flowers and bore rather leathery, smooth-margined leaves throughout the year. There were periods of relatively scant rainfall, but these were irregular and non-seasonal. The trees were related usually to modern broad-leaved trees, but in appearance they would have probably seemed as unfamiliar as the forests of Chile, New Zealand, or Australia seem to Northern Hemisphere visitors today. Conifers, such as members of the bald cypress and cedar families, may have preferred wetter areas, either near swamps or on the hilly terrain rising above the plains. Ferns were well represented in the flora, and plants of ancient lineage including ginkgos and cycad-like forms were present as well.

The dinosaurian assemblage is an interesting one, but skeletal materials are usually very incomplete. Forested environments were very widespread, and skeletons usually disintegrated on the forest floor; stream environments, where banks of sand can rapidly bury and preserve carcasses, were correspondingly less common. As a result, although about as many different varieties of dinosaurs are represented by some bony parts as in the older sediments of Dinosaur Provincial Park in Alberta (in both regions remains of nearly three dozen kinds of dinosaurs are preserved), only about eight are known from skeletal parts in Montana complete enough to give some idea of the body form of the animal.

In warm climates today, it is usual for big-game mammals to live in greater abundance in open woodlands or on the plains, while smaller mammals are relatively more common in densely forested habitants. It is tempting to believe that analogous environmental preferences account for the identification of at least 16 varieties of small dinosaurs ranging in weight from about 50 to 350 kilograms in the badlands of Fort Peck Reservoir. No other assemblage is known to contain so many small dinosaurs, most of which are known from teeth or a few isolated bones. Among the saurischians are two unnamed carnivores (one weighing perhaps 75 and the other several hundred kilograms), a scavenging dromaeosaur, a long-legged toothless form related closely to *Chirostenotes*, the large-brained, raptorial *Troodon*, two varieties of ostrich dinosaur (including a small *Ornithomimus* weighing about 50 kilograms and an undescribed large form weighing over 200 kilograms), and an unidentified animal perhaps related remotely to ostrich dinosaurs.

Among the small ornithischian herbivores were a small, fleet fabrosaur and a surprising variety of dome-headed

175

Stygimoloch:
detail of animal in forest,
Montana, 67 million years ago,
during late Cretaceous time

dinosaurs, or pachycephalosaurs, the skull-caps of which were evidently resistant to biochemical decay. Among these were individuals belonging to the round-headed *Stegoceras*, spiny-headed *Stygimoloch*, the large, round-headed *Pachycephalosaurus*, and two undescribed forms. Postcranial remains of *Stygimoloch* also occur in some abundance. By far the most common small herbivore was *Thescelosaurus*, which attained weights of 350 kilograms and lengths of over 3.5 metres. The animal possessed a small head, a short neck, small forelimbs, a barrel-shaped body, moderately long legs, and a long, broad tail. Perhaps it was semi-aquatic in the manner of modern capybaras and tapirs, and for this reason its skeletons are preserved frequently in stream-channel sands in a high degree of completeness. Finally, tail vertebrae of a small unidentified, possibly plant-eating (ornithischian) dinosaur are not uncommon. Taken as a group, these small dinosaurs were diverse enough to suggest the existence of a community of small carnivores and herbivores that was more or less independent ecologically of its larger relatives. Their interactions with each other and their forest habitat must have been fascinating, and it is unfortunate that they are so incompletely known.

Streams flowing through woodlands cut into their banks on the outside edges of meandering loops, and trees toppled into them. Abundant debris from the forests and sheltered environments provided a superb environment for turtles. Indeed, the stream-side habitats on the ancient floodplain hosted an abundance and variety of turtles probably exceeding that of any comparable modern habitat. Excluding microfossil materials, turtle bones are by far the most common fossils in the Fort Peck badlands. There were mollusc-eating, long-tailed baenids, fish-eating snapping turtles, large herbivorous dermatemydids, and many kinds of soft-shelled turtles, which then as now probably fed on small aquatic animals. Large, tortoise-like *Basileyms* specimens browsed in low vegetation growing on the stream banks and adjacent flatlands. The voice of the turtle was heard prominently throughout the lowland forests. More different kinds of mammals inhabited stream-side environments than did those of turtles, but they were much smaller and their teeth are much less frequently found. A greater variety of tree-dwelling multituberculates was present than in Dinosaur Provincial Park, and rare, ratlike omnivores (*Protungulatum*) lay close to the ancestry of modern herbivorous mammals. In addition to the small carnivorous dinosaurs that preyed on the small backboned animals, there

Screwpine proproots,
western Java

were monitor lizards present as large as the modern Komodo Dragon of Indonesia.

The dinosaurian megafauna has been termed the *Triceratops* fauna because of the abundance of large, horned dinosaurs in it. These animals were very large in comparison with other dinosaurs in the late Cretaceous of the northern plains, approaching 12 metric tonnes in weight and 8 metres in length. Few associated postcranial skeletons have been collected, but the three-horned skulls with their rigid neck shield resisted decay well and were more frequently buried. All certainly identified *Triceratops* specimens found in the Fort Peck fossil field were from fully adult specimens. Circumstances suggest that the animals nested elsewhere and that the midregions of the floodplain were areas in which migrating animals dispersed to feed. The specimens that have been recovered represent large and old individuals that died alone. Two kinds of *Triceratops* have been identified here: one had slender, recurved brow horns and a small, stubby nose horn. The former type is much more abundantly preserved than the latter, so these differences may not be caused by differences of sex.

Two other large horned dinosaurs were also present, isolated bones of which could be confused with those of

Triceratops horn core, weathering out of Hell Creek Formation, Hell Creek State Park

177

Triceratops:
detail of animal on a bayou mudflat,
Montana, 67 million years ago,
during late Cretaceous time

Triceratops. One had a long muzzle with a very rudimentary nose horn and nearly vertical brow horns.The proper name for this animal is uncertain, but it may be called *'Diceratops.'* The other large ceratopsian was *Torosaurus,* with a very long neck shield and relatively slender limbs. Remains of a tapir-sized small ceratopsian are not uncommon and have been considered juvenile specimens of *Triceratops.* They are from immature animals, but none attains the proportions of a half-grown *Triceratops,* and the brow horns project forward in a rather dissimilar curve, resembling that of cow horns. Where and what were the adults? Where were the half-grown individuals of *Triceratops*? All the ceratopsians so far collected at Fort Peck seem to be of the long-frilled chasmosaurine variety. No rhino-horned centrosaurine specimen has yet been collected, even though horned dinosaurs rival hadrosaurs in abundance in the fauna. Had the centrosaurines been replaced, or did they seek other habitats?

Hadrosaur remains are not as common as those of large horned dinosaurs in badlands south of Fort Peck Reservoir, but they appear somewhat more common in sediments exposed 45 kilometres to the east. The coastline at that time lay 400 kilometres further east, and it is doubtful that coastal effects could have increased the numbers of hadrosaurs. Ecological factors linked to a buried ridge in the Alberta dinosaur park were postulated to have caused a local abundance of hadrosaur remains. Along the Montana-Dakotas border, north of the rapidly growing delta in Wyoming, a ridge of low hills had also emerged briefly from the margin of the sea. It was much closer to the eastern edge of the Fort Peck fossil field. These hills were buried soon afterward by the alluvial plain supporting the *Triceratops* fauna. Yet for a few hundred thousand years they may have provided a firmer, possibly less heavily vegetated pathway for migrating hadrosaurs.

Skeletal materials of two kinds of the flatheaded hadrosaur *Edmontosaurus* have been collected from channel sands left behind by streams crossing the forested floodplain. The more common form closely resembled its counterpart in the coastal swamps of Alberta. Unlike in *Triceratops,* many specimens belonged to subadult individuals. One tiny skeleton was from a metre-long infant, which had only recently left its nest. The specimens probably represent attritional mortality from groups of animals in transit. *Edmontosaurus* skeletal fragments are also typically much more completely preserved than those of *Triceratops,* suggesting that the horned dinosaurs were not following stream courses to the extent that these hadrosaurs were. A second edmontosaur variety had a long, low head and a very broad, 'spoon-billed' beak. Others may be represented by isolated limb bones. A relatively short and heavy thigh-bone resembles that of tube-headed lambeosaurs. The presence of a very delicately proportioned hadrosaur is suggested by an arm bone. A

The 'Modern' Cretaceous

Tyrannosaurus and *Edmontosaurus*

LATE CRETACEOUS, ABOUT 67 MILLION YEARS AGO

A *Tyrannosaurus* brings down an *Edmontosaurus*,
as a giant pterosaur passes overhead,
in the sky over eastern Wyoming.

large femur was derived from the skeleton of an animal weighing well over 4 metric tonnes, much heavier than any other hadrosaur known from the Fort Peck region.

Other large dinosaurs are represented by equally incomplete material. The presence of the large armoured dinosaur *Ankylosaurus*, which was about 7 metres long and weighed over 4 metric tonnes, is documented by only one fragmentary skeleton. A few teeth and bones have been found pertaining to the armoured nodosaurs. A single tooth, said to have been found in these sediments, resembles those of early Cretaceous iguanodonts. And a peculiar ankle bone of a moderately large dinosaur closely resembles the same bone in bipedal, otherwise slothlike dinosaurs from China and Mongolia called segnosaurs.

The largest tyrannosaurids, and the largest land-dwelling carnivores known, are represented by a half-dozen skulls and skeletons in various stages of completeness. If one stands before a mounted skeleton, composed invariably of parts from different individuals, it is impossible not to reflect on the horrible power of these elephantine instruments of death. It would indeed be rather academic to choose between a lion and a polar bear in terms of which predator is more to be feared. Those who have some familiarity with both animals in the wild, when placed in the utterly abysmal situation of having to choose which hungry animal to confront, would possibly agree that the bear is more dangerous. These tyrannosaurs were two and one-half times longer than a large polar bear, and five times as massive. One of them was the famous *Tyrannosaurus*. The remains of a few equally large specimens have long been confused with those of *Tyrannosaurus*, but characters in the jaws and pelvis suggest that they belonged to a different animal. Both were active, muscular predators, and in all probability both were efficient hunters. Popular descriptions to the contrary, when these giants chose to attack, the herbivore was seldom the victor.

The dinosaurs of the Fort Peck fossil field show the effects of forest-dominated environments. Small dinosaurs were present in unusual diversity. However, at the opposite extreme many members of the dinosaurian megafauna were unusually large. Among the latter were giant tyrannosaurs, ankylosaurs, and ceratopsians. Their size may have reflected climatic conditions warmer than those a few million years previously, and an unusual abundance of food in their habitats. Within the general forest cover, were the stands of reedy vegetation thought to have been the preferred fodder of ceratopsians much more extensive than shrublands more suited for browsing hadrosaurs? Whatever the case, a much larger proportion of the dinosaurs

remained within the forested plain most of the year than crossed it in migrating groups.

To the Edges of the Plain

Some 700 kilometres south of Fort Peck Reservoir, and about 100 kilometres north of old Fort Laramie in eastern Wyoming, lies another fossil field where the presence of a dinosaur fauna dominated by *Triceratops* was discovered near the end of the last century. Its strata (Lance Formation, 760 metres thick) are exposed within the basin of Lance Creek, a tributary of the Cheyenne River. This region was closer to the ancient delta crossing northern Wyoming and received much more sediment. A greater number of streams crossed the woodlands, and the forest itself was not quite the same as in Montana. The leaves of the trees were smaller, and the hardwood forests were dotted with tropical conifers (araucarians). Clumps of palms would have probably provided the greatest visual contrast with more northerly forests.

Turtles seem not to have been as common as they were in Montana, although small mammals inhabiting the wooded wetlands were nearly identical in kind and in relative abundance. Other small animals inhabiting the streams and ponds closely resembled those of the modern Louisiana coastal plain. Lizards were present in much greater variety, but snakes were small and rare. Freshwater fishes and salamanders were similar to living forms, although present in less variety. Shore birds were also present, but bird life was much less varied and probably much less abundant as well. If, however, allowances were made for some impressive dinosaurs, smaller leaves on broad-leaved trees, and relatively fewer birds, a Cajun settler would have felt rather at home on the ancient Wyoming delta plain.

Remains of small dinosaurs have not been found in the abundance in which they occur in Montana. The porcine *Thescelosaurus* seems to have been relatively common, and of the pachycephalosaurs a spine of *Stygimoloch* and a cranial dome of *Pachycephalosaurus* have been collected. Among the large dinosaurs, *Triceratops* is represented by many skulls and a few fragmentary skeletons. The Wyoming skulls tend to be smaller than those from Montana. Perhaps the Wyoming range provided less fodder for *Triceratops* than was available further north, or perhaps it was more difficult for very large animals to avoid heat overload. Like the plates of *Stegosaurus*, the neck shield of *Triceratops* was grooved and channelled by the former passage of large blood vessels. The shield further resembles the half-spread ears of an African elephant. Perhaps, like that of its Jurassic predecessor, cer-

Tyrannosaur quarry,
in lower (Cretaceous) Scollard Formation,
near Huxley, Alberta

atopsian armour doubled as a heat exchanger. In any event, the size difference between the skulls of Wyoming and those of Montana specimens was real. As in the Montana locality, skulls belonging to the chasmosaurine *Torosaurus* occur much less commonly. Superb specimens of the flatheaded hadrosaurine *Edmontosaurus* have been discovered, and in the case of two spectacular 'mummies' the skeletons were found nearly completely enclosed in skin impressions. A jaw fragment of a very large hadrosaurine belonged to an animal weighing on the order of 8 metric tonnes. It was thus twice as large as any hadrosaur known in the *Triceratops* fauna. And assuredly preying on this giant biped, and on placid herds of other herbivores, were giant tyrannosaurs.

Scattered remains of *Triceratops* occur widely, from as far south as Denver, across Wyoming and the western Dakotas, to southern Saskatchewan. Over 500 kilometres north of the Fort Peck Reservoir, *Triceratops* bones occur also in the valley of the Red Deer River in Alberta, about mid-way between Calgary and Edmonton. Here its remains have been found in small area of badlands developed in floodplain deposits (lower part of the Scollard Formation, 40 metres thick), along with those of several dinosaurian varieties appearing with it in other areas. Among these are ostrich dinosaurs, *Thescelosaurus, Torosaurus, Ankylosaurus,* the large-brained *Troodon*, and a giant tyrannosaur. However, there seem to be fundamental differences between this small sample of dinosaurs and those of the *Triceratops* fauna further south.

Although *Triceratops* is less common here than at any other locality where it occurs, two of the skulls are from animals with heads 2.5 metres long, among the largest dinosaur skulls known. *Ankylosaurus* is not uncommon here, but no skeletal fragments of hadrosaurs have been collected. Beautifully complete skeletons of the small protoceratopsid, or hornless 'horned' dinosaur, *Leptoceratops* have been collected, and its remains outnumber those of any other dinosaur. These small dinosaurs weighed only about 200 kilograms. They resembled ceratopsids in general body form, but their legs were so long that they may have been capable of bipedal locomotion, and the tails were very high and narrow. They bore no horns on their skull and lacked even a rudimentary neck shield. *Leptoceratops* remains have been found elsewhere only in western Wyoming, in what were then the higher portions of the ancient plain. But both protoceratopsids and ankylosaurids are relatively abundant in sediments deposited in semi-arid steppes of central Asia, where ceratopsids are not surely known to occur and hadrosaurs are often rare.

Its dinosaur fauna clearly indicates that the central Albertan locality was close to the limit of the range of *Triceratops*. The broad floodplain ended against broken hills lining the front of Cordilleria some 200 kilometres further west. A long, narrow rainshadow extended north along the eastern margin of the mountains into southern Alberta, where dense forests could not grow. Caliche formed in the soil, and the sparse vegetation was dominated by primitive, drought-resistant conifers. And, as in central Asia, tiny insect-eating mammals probably flourished in the scrublands. Even in the *Triceratops* locality their remains nearly equal those of primitive opossums and archaic seed-eaters in abundance, whereas in the wet, lowland forests they were very rare.

Passage to the Northwest

The great northern forests of Canada cover the plains beyond central Alberta, and the northern limit of *Triceratops* is unknown. Dense stands of black spruce, followed by subarctic taiga, blanket a narrowing corridor of ancient sediments into the delta of the modern Mackenzie River. Exposures are few and limited to the banks of arctic streams. A few tantalizing remains of northern contemporaries of *Triceratops* have been discovered. A bone from the skull of a medium-sized horned dinosaur was collected from sediments deposited on a large alluvial fan (lower part of the Summit Creek Formation, about 100 metres thick) spreading from the Mackenzie Mountains west of Great Bear Lake in the Northwest Territories. The locality was located just north of the Arctic Circle (67 degrees north) during late Cretaceous time. At a late Cretaceous latitude of 70 degrees north in northern Yukon, bones from a half-grown hadrosaur were discovered in sediments deposited on a swampy lowland

river plain (Bonnett Plume Formation, about 1200 metres thick). Finally, tyrannosaur teeth and remains of horned dinosaurs and of juvenile and adult hadrosaurs have been collected near the mouth of the Colville River on the north slope of Alaska. The site was at a latitude of 80 degrees north, only a thousand kilometres from the ancient pole.

From the point of view of an animal, polar regions then represented a food source not unlike the broad-leaved forests of eastern North America today. During part of the year, these forests are bare and plant growth ceases. The long, warm days of spring produce a flush of plant growth. Small, non-migrating animals begin their life cycles anew, feeding on nutritious, tender shoots. Many birds leave tropical ecosystems and fly north to take advantage of the great burst of food that comes with the northern spring. In late Cretaceous times, plant growth also burgeoned with the onset of the summer day. A narrow coastal plain extended from central Alberta to northern Alaska, and fossil pollen within it indicates that the vegetation growing on it was similar from midlatitudes to the ancient Arctic Circle. It was virtually a highway to the north. Hadrosaurs did migrate and could probably travel overland for distances exceeding the annual migration of many modern birds. Some of the horned dinosaurs herded as well, and long-legged tyrannosaurs undoubtedly followed the migrating herbivores. In all probability the arctic was filled with activity during the midsummer day. It fell silent beneath the polar night, when the trees were bare, plant growth had ceased, and all the dinosaurs had left.

With Alaska so near the pole, a dinosaur on the north slope could effectively travel south in two directions: one along the North American 'highway' to the *Triceratops* forest, the other into China. The land route to Asia had not always been open. During early Cretaceous time, about 125 million years ago, sauropods, stegosaurs, and small, bipedal archaic ceratopsoids (psittacosaurs) inhabited regions around the interior lakes and steppes of what has become the Gobi Desert. These animals did not closely resemble their contemporaries in North America. A land link seems to have been established about 100 million years ago, when small carnivorous dinosaurs, ostrich dinosaurs, dome-headed dinosaurs, and primitive hadrosaurs and ceratopsians evidently formed a common stock from which evolved later Cretaceous dinosaurs on both continents. By the time of *Triceratops*, dinosaurian stocks in both areas generally resembled each other, and these similarities had probably been enhanced by the establishment of a Cordillerian land link a dozen or so millions of years previously.

Remember the Alamosaurus?

Accompanying the rapid expansion of the forested plain inhabited by *Triceratops*, broken chains of 'island hills' and lakes formed, curving from the mountainous highlands to the southwest through Wyoming and then south through central Colorado and New Mexico. A large, inland basin thus formed between highlands in Idaho and western Utah and the broken terrain to the east. *Triceratops*, or a similarly large-horned dinosaur, entered this basin in at least one area near its northern limit. Here, in southwestern Wyoming, a ceratopsian lower jaw was found in proximity to cobblestones washed by a large river from the western highlands (lower part of Evanston Formation, about 300 metres thick). In the same region, and in deposits of the same river system, material of a large sauropod has also been collected – the northernmost occurrence of long-necked, long-tailed brontosaurian dinosaurs. The remains of these animals are unknown from sediments deposited in the southern United States immediately before the expansion of the *Triceratops* plain.

Two hundred kilometres south of Salt Lake City, a partial skeleton of a large sauropod was collected from stream sands in a series of river and lake deposits accumulating west of hilly terrain in eastern Utah (lower part of North Horn Formation, 335 metres thick). Called *Alamosaurus*, it belonged to a group of sauropods characterized by a large, ball-like cap on the rear surface of the tail vertebrae that fits into a cup on the following vertebra. These dinosaurs, or titanosaurs, are known best from the southern continents, although their remains have been identified in Eurasia as well. Because they evidently did not enter northern North America, the land-bridge to Asia was not accessible to them. Titanosaurs must have entered from South America along a volcanic bridge spanning the Caribbean Sea. Unfortunately, the most complete *Alamosaurus* specimen in Utah is very fragmentary, but its nearly 3-metre-long forelimb is compatible with giraffoid, tree-browsing habits. It was not a giant sauropod, for it probably weighed only between 25 and 30 metric tonnes. By way of comparison, one very large *Triceratops* specimen weighed 19 metric tonnes. This horned dinosaur had probably attained a relatively very advanced age, for its bones show evidence of disease indicative of declining health in a very old animal. However, *Alamosaurus* was the largest dinosaur known in North America during its time.

Horned dinosaurs seem to have been common in the region of the *Alamosaurus* locality in central Utah. Fragmentary skulls belonged apparently to the long-frilled *Torosaurus* and *Pentaceratops*. Hadrosaur remains are uncommon and incomplete. Isolated tyrannosaur bones were derived from medium-sized animals. Fragments of dinosaur eggshells are locally very abundant and belong to many different varieties. One of these is represented by relatively small but well-preserved eggs 16 centimetres long and originally weighing about half a kilogram. Small dinosaurs that may have been egg predators include ostrich dinosaurs and a long-legged, toothless animal related to *Chirostenotes*. Freshwater turtles were present, as well as the large tortoise-like *Basilemys*, but the most interesting small reptiles were abundant herbivorous lizards (*Polyglyphanodon*) belonging to an extinct group allied to the skinks. These animals had large heads and short necks and averaged a metre in length. Close relatives inhabited the region of the Gobi Desert in central Asia at this time, and the animals may have avoided lowland deltaic environments.

Grey fluviodeltaic sediments near Bryce Canyon National Park in south-central Utah (Kaiparowits Formation, 800 metres thick) have not yielded *Alamosaurus* bones, although the turtles and small carnivorous dinosaurs were apparently identical to those from the central part of the state. These badlands have not been well explored but have so far produced a well-preserved *Ornithomimus* skeletal fragment and a piece of the trombone-like crest of the hadrosaur *Parasaurolophus*. Horned dinosaur and nodosaur remains have also been identified.

In northwestern New Mexico, 350 kilometres away, numerous *Alamosaurus* bones and a few skeletal fragments indicate that it was locally one of the most common of dinosaurs. Its remains occur in sands and gravels (Naashoibito Formation, 20 to 50 metres thick) that were eroded from volcanic highlands in southern Colorado. The alluvial fans were covered with conifer-dominated forests, although palms were present as well. Most of the dinosaur skeletons were broken apart by the rapidly flowing streams, and bones were usually scattered and water-worn. Ceratopsians were once again represented by *Torosaurus*, which seems to have rivalled *Alamosaurus* in abundance, and by the less common *Pentaceratops*. The flatheaded hadrosaur *Kritosaurus*, which possessed a reflexed snout recalling a Roman nose, occurs in the fauna, as it had in the older faunas of New Mexico and Alberta. More significant, however, its presence has been documented in Argentina in late Cretaceous strata. The land corridor across the Caribbean, and tropical climates in general, were as congenial to *Kritosaurus* as they were to *Alamosaurus*. While the sauropods were extending their range across the isthmus to the north, the kritosaurs were moving south through the same region.

Completing the inventory of dinosaurs from the *Alamosaurus* fauna in New Mexico were representatives of the small dinosaurs with large brains (*Troodon*), medium-sized tyrannosaurs, and armoured, clubless nodosaurs. Turtles were abundant and varied, although fewer fishes inhabited the clear, rapidly flowing streams than occurred in contemporaneous lowland deltas.

The *Triceratops* fauna extended north-south over at least 1500 kilometres between central Alberta and central Colorado during late Cretaceous time. The *Alamosaurus* fauna, its southern contemporary, also extended over 1500 kilometres in the same direction, from western Wyoming to Big Bend National Park in western Texas. In the Big Bend, meandering streams deposited varicoloured sands and clays on a hot subtropical plain 400 kilometres from the edge of the interior sea (Javelina Formation, 240 metres thick). There was enough moisture to support large trees. Few dinosaur specimens have been collected from these sediments. Sauropods, including *Alamosaurus*, were present, as well as a rather small specimen of the long-frilled horned dinosaur *Torosaurus*. A few bones suggest the presence of a smaller relative of the northern *Tyrannosaurus* and a flatheaded hadrosaur allied to *Edmontosaurus*. Small carnivorous dinosaurs are represented by foot bones.

The most famous fossils from the western Texas sediments are bones from the wing of a gigantic pterosaur. Remains of 12 individuals of similar but smaller specimens were discovered from a second site in the same strata. Although several differing estimates have been made, it seems likely that the large animal had a wing-span of about 15 metres, the smaller ones, about 5 metres. The pterosaur has been named *Quetzalcoatlus*, and what is known of its anatomy has recently inspired the construction of flying scale models. Images of *Quetzalcoatlus* in life defy if not defeat the imagination. It is useful to measure off a distance of 15 metres in order to appreciate the size of the animal. Pterosaurs probably took off by running on their hind legs into the wind and flapping their wings. This animal stood perhaps twice as high as a human and weighed about 100 kilograms. Its arms were more muscular than the legs of a very powerful man. What colour were its eyes? Was its body black like that of many modern tropical birds of prey?

Quetzalcoatlus surely required a large quantity of food. There were few lakes, and fewer still that did not periodically dry out, so that fishes were unavailable as a dependable food source. There is no indication that the region supported dinosaur rookeries, and with them a supply of relatively helpless young animals. The giant pterosaur's wings would be at some risk from the struggles of medium-sized herbivores, which were adapted in some manner to resist the attacks of carnivorous dinosaurs. Perhaps it was in the habit of overflying the region en route to richer feeding grounds, although the occurrence of 12 smaller and perhaps immature individuals in one site does seem to suggest a stable link to the local ecosystem. Perhaps the animals were able to spot large dying or dead dinosaurs over a broad area of terrain and thus fed abundantly on carrion.

The California Coast

During the time of *Triceratops*, California was, as it still is, earthquake country. Great ruptures in the Earth's crust intersected with the shoreline, and coral atolls originally far away in the tropical Pacific were carried by moving crustal plates into the coastal ranges. Basin and range topography had not yet formed across Nevada and western Utah, and the consequent expansion had not yet pushed the Pacific rim of the United States as far west as it now lies. Climates along the California coast today are dominated by an alternation of rainy and dry seasons, although fossil woods suggest that rainfall then, though abundant, was evenly distributed throughout the year.

The rains washed dark silts from rugged coastal ranges into a deep trough. There they accumulated in submarine fans (Moreno Formation, 575 metres thick) which have subsequently been crumpled and transformed into the Moreno Hills, 75 kilometres west of Fresno. Wood, pollen, and spores were buried in the silts, together with the bones of marine reptiles, fishes, and, more rarely, dinosaurs. The latter apparently belonged to two kinds of flatheaded hadrosaurs, one with a cranial spike (*Saurolophus*) and another without head ornamentation. The plant remains indicate that nearby lands inhabited by the dinosaurs were densely covered with a highly diverse, subtropical-to-tropical forest. Conifers adapted to warm climates were a major element, but the forests were dominated by trees belonging to the more primitive living groups of flowering plants that also prefer the warmer regions of the Earth. To the south, between Los Angeles and San Diego, a few fragments of hadrosaur bones have been discovered in marine deposits the same age as those west of Fresno.

The Delaware Valley

Between 75 and 65 million years ago, as Cordilleria was transformed into the Canadian Rockies and the Pacific islands

Javelina Formation,
exposed in Dawson Creek valley,
Big Bend National Park,
near site of wing fragment
of giant pterosaur (*Quetzalcoatlus*)

Sauropod quarry (TOP),
pterosaur quarry (BOTTOM),
both in Javelina Formation,
Tornillo Flat, Big Bend

Scavenger storks,
roosting in tree,
Luangwa Reserve, Zambia

merged with the California coast, the western half of North America was destabilized. Rugged ranges of soft strata were thrust up, the crust was torn by huge rupture zones, and vast basins sank beneath the growing weight of enormous alluvial fans. Beyond the midcontinental sea, the eastern half of North America was relatively stable, and there were no great alluvial fans forming that could preserve a record of terrestrial life. Instead, a very thin veneer of near-shore marine sediment lapped onto the old, eroded core of the Appalachian highlands. The modern coastal plains extending from the region of New York, around that of Atlanta, and into the Mississippi Valley in southern Illinois, was covered by the sea. The Great Lakes region, eastern Ontario, and Quebec constituted an ancient shield area bordering the Appalachians on which the land-forms were consolidated and well vegetated. The clear rivers draining them took little sediment to the sea. Dinosaurs probably flourished in the Appalachian region, but their record could be preserved only in sediments. Sedimentary deposits of this time are meagre, as is our knowledge of eastern dinosaurs.

As has been noted, eastern forests were growing under warm, seasonless, rather dry climates. Lowland forests were dominated by short, small-leaved flowering trees, among which ancestral walnuts were particularly common. In well-watered areas, tall conifers grew in some abundance, perhaps accompanied by large, flowering trees bearing leaves like those of eucalyptus. The Gulf Stream already flowed north off the Atlantic coast, for forests growing in western Greenland (at a latitude comparable to that of modern Montreal) closely resembled those of the southern Appalachians. South of James Bay in Ontario, forests dominated by primitive walnuts blanketed what may have been a sugarloaf terrain similar to areas with spectacular limestone peaks in Brazil and southern China today. Blade-needled podocarps flourished with araucarias and ferns, along the ridges of the Appalachians, above the warmer and drier bushlands near the sea.

A narrow band of near-shore marine sediments extends for 300 kilometres from near New York City, past Philadelphia and Baltimore, to Washington, DC. These sediments were deposited on the floor of a shallow, clear sea and are typically less than a hundred metres thick. In them are preserved shells of marine organisms and, in some abundance, bones of fishes and marine reptiles. The dense and compact teeth of sharks of all sizes are especially common. Very rarely the sediments have also yielded dinosaur bones, carried by tidal rivers to the edge of the ocean. These occurrences are as valuable as they are infrequent, for they embody the totality of our knowledge of late Cretaceous dinosaurs in northeastern North America. The dinosaurs so far identified are probably not representative of dinosaur populations inhabiting the subcontinent, for only those living close to sea would have left skeletal parts behind that could easily be swept into marine environments.

One of the two partial skeletons of dinosaurs so far discovered in the region was collected in 1858 a few kilometres east of Philadelphia in New Jersey. The only known carcass of the flatheaded hadrosaur *Hadrosaurus* came to rest in sediments deposited beneath a shallow bay (Woodbury Formation, 8 metres thick) about 75 million years ago. The animal had been rather small, weighing a little over $1\frac{1}{2}$ metric tonnes. The find, however, was of international significance, for it was the first relatively complete dinosaur skeleton to be collected and it demonstrated that at least some dinosaurs were bipeds.

The second partial skeleton belonged to the carnivore *Dryptosaurus* and represented an adult weighing only about three-quarters of a metric tonne. This specimen was collected in 1866 from shallow marine greensands (Navesink Formation, 12 metres thick), deposited about 65 million years ago within a few kilometres of the *Hadrosaurus* locality. It too was a biped and differed from tyrannosaurs in having more slender legs, longer arms, and much larger and more recurved claws on its hands. The animal was probably not closely related to tyrannosaurs. The sediments have yielded a relatively large variety of dinosaur remains, including those of ornithomimids and of flatheaded and tubeheaded hadrosaurs. Strata of the same age, deposited within a shallow bay in the eastern suburbs of Washington, DC (Severn Formation, 6 metres thick), have yielded bones of hadrosaurs and a moderately large ornithomimid.

Another important dinosaur locality was discovered recently a few kilometres east of Trenton, in sediments left about 75 million years ago beneath the mouth of an estuary (Marshalltown Formation, 8 metres thick). The bones are badly abraded, probably as a result of storm-generated currents, and were mixed with plant debris from the nearby land. They have been identified tentatively as belonging to dinosaurs allied to *Dryptosaurus* and two varieties of flatheaded hadrosaurs. Occurring with them are tail vertebrae of a large flatheaded hadrosaur called *Hypsibema*, formerly considered to have been a sauropod. The giant 'dinosaur-eating' crocodile *Deinosuchus* was also present. Ornithomimid and hadrosaur bones have been collected in northern Delaware from sediments deposited at the same time and in a similar environment.

Moreno Formation,
exposed in Panoche Hills,
west of Fresno, California

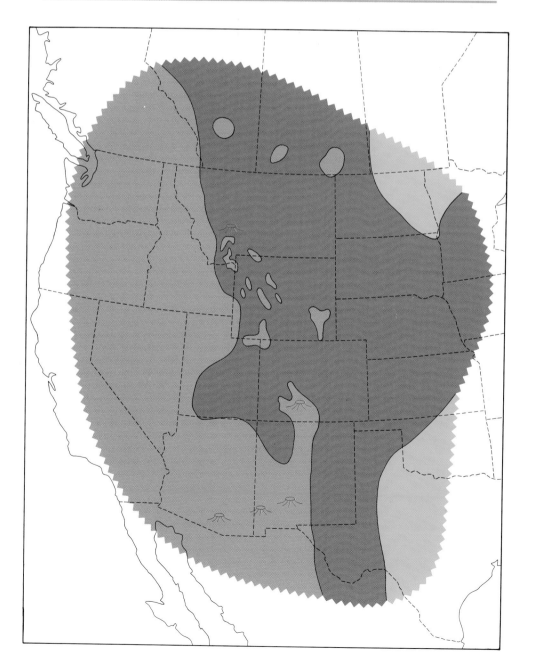

About 65 million years ago

In western North America,
interior seaway again separates
into northern and southern arms.
A broad, fertile plain dotted
with highlands follows eastward-
retreating arms of the sea.

Dryptosaurus

LATE CRETACEOUS, ABOUT 67 MILLION YEARS AGO

Dryptosauri court beneath setting Sun, along estuary in
New Jersey. Several varieties of now-extinct birds inhabit region;
large tree is related to bombax family, now tropical.

Dixie Dinosaurs

Probably as a result of rivers descending along the axis of the Appalachian ridges, a greater thickness of alluvium accumulated on lowlands bordering the sea in the southeastern United States. Modern soils are very deep, and these sediments are usually exposed only in freshly eroded stream banks. Because the sedimentary record is more complete than in the Delaware Valley, and because the area was then located closer to the equatorial zone, a search for new dinosaurian occurrences should be encouraged.

Five hundred kilometres south of Washington, DC, along the banks of streams in the Cape Fear basin of southern North Carolina, silts can be seen that were deposited 75 million years ago in estuaries littered with logs and plant material (Black Creek Formation, 120 metres thick). Waterworn bones and teeth from 1-metric-tonne carnivorous dinosaurs, 2-metric-tonne flatheaded hadrosaurs, and giant crocodiles (*Deinosuchus*) have been excavated here. Remains of ostrich dinosaurs of moderate size have also been collected, as well as several vertebrae from tails of the otherwise unknown large dinosaur *Hypsibema*.

A belt of late Cretaceous marine strata curves from Georgia 400 kilometres across central Alabama into northeastern Mississippi. South of Columbus, Georgia, in the valley of the Chattahoochee River, lagoonal sediments with abundant plant remains accumulated about 80 million years ago behind barrier sand-bars (Blufftown Formation, 180 metres thick). Bones of small carnivorous dinosaurs and hadrosaurs occur here as well, with abundant remains of the huge, 15-metre-long crocodile *Deinosuchus*. Seventy-five kilometres west of Montgomery, in the Black Prairie region of Alabama, bones of a carnivorous dinosaur, flatheaded hadrosaurs, and a nodosaur were recovered from open-water marine chalks (Mooreville Chalk, 75 metres thick) of the same age. In Mississippi, north of Columbus, near-shore marine sands (Eutaw Formation, 90 metres thick)

have yielded the bones of hadrosaurs and a small tyrannosaur. Finally, in southeastern Missouri, about 50 kilometres east of the confluence of the Mississippi and Ohio rivers, an articulated series of tail vertebrae of *Hysibema* was taken from a private water-well. Regrettably, the rest of the skeleton was never collected, and the appearance of the dinosaur remains unknown. The specimen was buried in near-shore sands about 70 million years ago (McNairy Formation, 95 metres thick).

Because eastern North America was separated from the west by a seaway during much of late Cretaceous time, it would be reasonable to expect that the effects of geographical isolation would be apparent in the dinosaur faunas of the Appalachian region. These would be enhanced by a different flora dominated by walnut-like trees and by the absence of great north-south migration routes. So far no horned dinosaur remains have been identified in the east, though ceratopsians are very rarely found in marine sediments. The regional abundance of giant crocodiles is probably the result of environmental preferences, for they also occurred west of the seaway. *Dryptosaurus* is known so far only from a single fragmentary skeleton from New Jersey, although it could possibly be represented in the west by other, unrecognized skeletal parts. The enigmatic hadrosaur from Missouri, *Hypsibema*, seems a good candidate for the typical dinosaur of eastern North America but unfortunately is known only from 'the part that went over the fence last.'

Epilogue

The dinosaurs of the *Triceratops* fauna and contemporaneous dinosaur faunas in North America and around the world never evolved any further. No unambiguous indication that any dinosaurs were alive on this planet at some subsequent time has ever been found. Their story ends here.

The extinction of the dinosaurs

65 MILLION YEARS AGO

One second before impact,
comet passes through
Earth's atmosphere.

About 10 seconds after impact, a plume of vaporized rock and water rises 100 kilometres through the atmosphere. Within about 20 minutes, the shock wave will kill the saurolophine hadrosaurs in the foreground.

The
Extinction
of the
Dinosaurs

Notice to the Reader

On the basis of what you have read so far, you will not be able to solve the mystery of the extinction of the dinosaurs. This is because there *are* no clues in the dinosaurian record; they lie elsewhere. It is highly probable that the Earth was struck by a cometary body or an asteroid about 65 million years ago and that the environmental disturbances generated by the impact were too great for the dinosaurs to endure. Other explanations for the extinction of the dinosaurs are based on a more limited range of information. Some postulate that the dinosaurs were slowly declining to extinction, that dinosaurs survived as birds and hence never became extinct, that the survival of many kinds of organisms through the time during which the dinosaurs became extinct rules out any unusual environmental disturbances, or that existing evidence is too incomplete and too ambiguous to show how the dinosaurs became extinct. These hypotheses do not explain circumstances surrounding the extinction of the dinosaurs very well. Some, however, have been useful, for they have stimulated the search for new information which in turn rendered them unlikely. Thus the alternatives were narrowed.

The Impact of an Extraterrestrial Object . . .

Evidence that the impact of an extraterrestrial object coincided with massive extinctions is very strong. Well over half of the different varieties of organisms on Earth became extinct with the dinosaurs. By consulting the record of tiny and abundant microfossils, a thin layer of sediment can typically be identified that was deposited at the time of the extinction (the 'Cretaceous-Tertiary boundary'). This tiny stratum occurs world-wide and contains chemical and physical evidence of extraterrestrial material.

One group of metals is attracted to iron and occurs much more abundantly at the Earth's core than in its crust. It includes nickel, gold, platinum, osmium, and iridium. These metals are also abundant in meteorites. The quantity of these metals contained in the thin, global extinction stratum corresponds to the quantity in a meteorite or cometary nucleus measuring 10 kilometres in diameter. A 10-kilometre object would strike the Earth with a force about 10,000 times the energy released by the simultaneous detonation of all existing nuclear weapons. The object would be vapourized, and a cloud of condensing meteoritic and target materials would be blown around the world. Tiny spherules measuring a third of a millimetre in diameter have been identified around the world in the extinction stratum, together with evidence of glass and deformed crystals that were blown from the impact site. A crater measuring over 150 kilometres in diameter would have been created. A crater-like excavation measuring 300 kilometres in diameter on the floor of the Indian Ocean north of Madagascar could represent the impact site, although the possibility remains speculative. Fluffy particles of carbon similar to smoke particles from burning wood occur in spectacular abundance in the extinction stratum. They have been interpreted as evidence of enormous wildfires that burned as much vegetation as is now growing in tropical areas of the Earth.

Everything within a thousand kilometres or more of the crater would be crushed beneath the weight of the ejecta apron, ignited by the fireball, or torn apart by the shock wave transmitted through the air. Gigantic slumps in once loosely consolidated sediments on the continental shelf off

Cottonwood tree, near Ojo Alamo Spring
and abandoned store, San Juan Basin,
southeast of Farmington, New Mexico

Kenya and Somalia may have collapsed at this time, and enormous outpourings of lava on the Indian subcontinent may have been triggered. But these effects would have been local and could not have produced a global disaster. However, the tremendous energy released in the impact would dwarf the restraining effects of gravity and air pressure. A great quantity of dust would be blown up the transient hole in the atmosphere created by the high-velocity extraterrestrial object and rapidly envelop the Earth. The Sun's rays could not penetrate such a dust veil, and for weeks the surface of the Earth would remain darker than on a moonless night. Temperatures would fall rapidly, particularly in the interior of continents far from the warming waters of the sea. Atmospheric nitrogen would burn in the tremendous heat of the impact zone, to be washed out of the atmosphere around the Earth in acid rains. The biological effects of these stresses are becoming understood with some precision. Imagine how a garden would respond to many weeks of continuous darkness, near-freezing temperatures, and acid drizzles. How did the dinosaurian world respond?

Cretaceous-Tertiary boundary,
near Ojo Alamo Spring, San Juan Basin.
FOREGROUND: Upper Shale Member of Fruitland Formation, overlain
by yellowish sandstones of Naashoibito Formation, which
preserves bones of *Alamosaurus*.   ON HORIZON: Light sandstones of
Ojo Alamo Sandstone, which contain fossil Tertiary tree trunks

Frond from a palm that grew shortly
after the dinosaurs became extinct,
collected from sandstone in Raton Formation,
above boundary clay in Berwind Canyon,
north of Trinidad, Colorado

Boundary clay (detail), Berwind Canyon

Cretaceous-Tertiary boundary, in Berwind Canyon.
The boundary is marked by thin white line of
fallout, which dropped into clays of
an ephemeral swamp, visible above massive
sandstone at base of exposure.

Cretaceous-Tertiary boundary,
Red Deer River valley, east of Huxley, Alberta.
Fallout material occurs at base of Nevis Coal,
within Scollard Formation.

Detail of facing picture

An Odyssey in Time

Cretaceous-Tertiary boundary, north of
Jordan, Montana. Boundary occurs at thin
fallout horizon, just below dark horizon
(z coal), in upper third of photo.
Below coal: Hell Creek Formation; above:
early Tertiary Tullock Formation

. . . in the Sea

The mysterious brevity and world-wide extent of the extinctions terminating the age of reptiles were first observed in sediments deposited on the floor of the sea. Shells and platelets of floating marine algae and single-celled animals, measuring respectively 0.004 and 0.04 millimetre in diameter, are preserved in astronomically high numbers in chalk. Two-thirds of all species simultaneously became extinct, and their small size and great numbers made it possible to locate the time of the extinction within a single sedimentary stratum. This stratum is quite peculiar and has been studied in many locations, both from sediments now exposed on land and in cores still on bottom of the oceans. In it are preserved a basically uniform sequence of events.

1: Chalky oozes accumulating on the sea floor during the last years of the age of reptiles show no sign of impending ecological changes.

2: Chalk deposition ceases abruptly, and the chalks are covered by a light brown layer about half the width of a pencil in thickness. The layer has a high metallic content, and the metals occur in the same proportions as they do in meteorites. The layer also contains particles of shocked minerals, condensed droplets of once-liquefied rock, and decomposed shards of glass. These materials settled out from an ejecta cloud that had been blown over the atmosphere and around the world. The fallout was undisturbed by the activity of living things, and its chemistry suggests that there was little oxygen dissolved in waters near the sea floor. It was a time when the sea was essentially dead.

3: The brown layer is followed by a layer of grey clay, formed from fine clay particles eroded from the land. Some of this material is polluted with meteoritic debris that had fallen originally on land. The grey clay, representing a period of about 10,000 years, gradually becomes lighter from the swelling infall of carbonate from the shells of marine microorganisms. A few varieties of Cretaceous algae survived, mingled with periodic and sudden expansions in the populations of other ecologically opportunistic algae. The Cretaceous survivors finally completely vanish, as a result of unknown causes.

4: An ecological balance is achieved once again, with the evolution of new communities of planktonic organisms. Chalky oozes again form on the sea floor from the rain of microscopic shells of creatures living in surface waters. Very little land-derived clay is mixed with the chalk, but the particles are larger and suggest an increase in the rate of flow of rivers from the land. The surviving planktonic algae evolve rapidly and are able to produce as much plant food as before the extinctions, after the passage of about one and one-half million years.

Micro-organisms living close to the surface of the oceans are abundant and can reproduce at very high rates. They form the ecological base of the food chain in the sea. The decimation of populations of micro-organisms coinciding with the world-wide fallout of meteoritic dust almost certainly produced mass starvation. A greater proportion of bottom-dwelling organisms became extinct than of floating micro-organisms, and a greater proportion still of free-swimming marine creatures. There were other peculiarities in the pattern of oceanic extinctions. Previous abundance was of little value to survival, although groups that did not range widely around the globe more often did not survive. In the oceans, the age of reptiles ended with the extinction of the mosasaurs, plesiosaurs, toothed birds, and many kinds of bony fishes. But some backboned animals were able to endure the crisis, including many sharklike fishes and some marine turtles and crocodiles. It was a catastrophe, not an annihilation.

. . . and on the Land

The brevity of the extinctions is strikingly apparent on land as well. Long after the end of the age of reptiles, sedimentary sequences deposited during that time were turned on edge, and they are now eroding in the eastern foothills of the Rocky Mountains. They are typically 2 to 3 kilometres thick, and nearly three-quarters of this thickness accumulated as Cordilleria and other oceanic blocks collided with North America during Cretaceous time. Pollen and spores shed from land plants are similar in size to the shells of microscopic animals in the sea. With them, an extinction horizon can be located as precisely in terrestrial sediments as it is in marine chalks. The horizon can be traced from fossil floodplains, through fossil deltas, and into fossil bays, where it is found to coincide with the extinction horizon in the ocean. The extinction layer on land also is composed of a pencil-width thickness of fallout enriched in extraterrestrial metals. The fallout has been identified immediately on top of dinosaurian-age sediments between New Mexico and Alberta, over a distance of at least two thousand kilometres. Thus, the record of the development of the age of reptiles is measured in kilometres, while the record of its decline is measured in centimetres. In the latter record is preserved a complementary and basically uniform sequence of events.

1: The youngest sediments of the age of reptiles contain abundant and diversified pollen grains of flowering plants. Bits of plant debris show that cuticles of the generally small evergreen leaves were thick and adapted to warm and rather dry climates. Soils were well developed and non-acidic.

2: The soils abruptly merge into a brownish layer with a high metallic content, particles of shocked minerals, condensed droplets of once-liquefied rock, and decomposed shards of glass. Sediments show evidence of the passages of gigantic tidal waves across shallow seas to devastate nearby lowlands. Great quantities of soot document the ignition of firestorms of subcontinental proportions. Smoke intensified the blackness of a sky which was illuminated from below by stygian sheets of flame spreading through a dying world. More than half of the species of land plants become extinct, particularly those unable to shed their leaves and adapted to warm climatic conditions. The extinction pattern is compatible with a temperature decline to near-freezing temperatures for one to two months.

3: A barren interval a few millimetres thick follows, containing no evidence of growing vegetation. Then fern spores greatly increased in abundance, as ferns spread through the impoverished soils and charred remains of Cretaceous forests.

4: For reasons that are not understood, rainfall increased. Floodplains changed into peatbogs. Rivers draining the broad western flatlands increased in size and rate of flow. They cut deeply into older deposits, often mixing in their channel sands remains of organisms that lived before the extinction with those that lived after it. Organic acids from the peats penetrated old, pre-extinction soil, dissolving previously buried bones and further confusing the record.

5: Spreading from a few protected areas, trees followed the ferns back across the devastated and flooded plains. Bald-cypress colonies established themselves in peatbogs. Although climates were as warm as before, the surviving flowering plants belonged usually to varieties with large, thin leaves that were shed during periods of short day length. New forests formed from survivors of native pre-extinction forests; none had immigrated from other continents. A new plant-landscape equilibrium came into being that would last for several millions of years.

It is remarkable that the dinosaurs, which had been present on all continents, apparently also became extinct everywhere. We must conclude that there was no haven any-

where on the planet that permitted the survival of a single dinosaurian species through the crisis. In itself, the annihilation of the dinosaurs is one of the most amazing events ever to have occurred in the Earth's history.

After this stunning event, having weathered an ecological trauma that had been unequalled for over a hundred million years, survivors from the age of reptiles awkwardly merged into relatively simpler and more crudely balanced ecosystems. The former variety and abundance of big animals were suddenly gone. A few aquatic turtles, some primitive bony fishes, and a salamander may have attained weights approaching 25 kilograms shortly after the extinctions, but could these overgrown and clumsy creatures balance the loss of a megafauna? The giant estuarine crocodiles were gone, pterosaurs no longer soared in the skies, and the ranks of lizards, of opossums, and of seed-eating mammals were decimated. A plague of ratlike mammals spread over the land, infesting everything and indiscriminately devouring eggs, insects, fallen fruit, and even carrion. Majesty, proportion, and beauty were diminished in a new world that did not seem particularly brave. The warm, moist air was permeated with loss.

There were oddities in the pattern of extinction and survival. Sharklike fishes virtually disappeared from freshwater streams, although their relatives fared better in the seas. The crisis did not disturb the continuity of the history of frogs and salamanders, or that of the gavial-like champsosaurs, which died out much later. On the whole, life in the freshwater streams differed little. Opossums, which nearly became extinct in North America, survived in abundance in South America.

In one sense, it was still an age of reptiles. In North Dakota, some 5 million years after the extinctions, streams gently flowing through bald cypress swamps to a nearby sun-drenched sea supported large populations of 4-metre-long crocodiles (*Leidyosuchus*). Accompanying them were large champsosaurs (*Champsosaurus, Simoedosaurus*) and small alligators (*Wannaganosuchus*). There were at least five kinds of aquatic turtles, including snapping (*Protochelydra*) and soft-shelled varieties (*Trionyx*). These reptiles were by far the largest animals living in the wetlands. But when old animals died, their bones were often gnawed by hyperactive, omnivorous mammals. The period of reptilian ascendancy would soon end.

Genesis of an Apocalypse

The end of the dinosaurian era may have begun this way.

A vast number of protocomets orbit the Sun far beyond the outermost planet, Pluto. One of these balls of stone and ice was slightly pulled from its orbit, perhaps by the gravitational tug of a small star that passed nearby in the course of a huge and elliptical trajectory around the Sun. When the protocomet began its drift toward the inner solar system, the Sun appeared as a bright star about one-fifth of a light-year away. The time was about 350,000 years before the extinction of the dinosaurs. Another 16 metres of alluvium of Cretaceous age would be deposited on the floodplain in Montana, and another 6 metres of Cretaceous chalk on warm continental shelves. Very slowly, the velocity of the protocomet increased.

It would have seemed as if nothing important was happening. About 26 years before the end of the Cretaceous the protocomet crossed the orbit of Pluto, moving at a velocity of 5 kilometres per second. Five months before the end of the Cretaceous it crossed the asteroid belt, and its speed had increased to 18 kilometres per second. When it approached the orbit of Mars the end of the Cretaceous lay 33 days into the future. It developed a luminous head and long, filamentous tail and was now visible from the surface of the Earth. No dinosaur responded. When the comet crossed the orbit of the Moon it was moving at a velocity of 30 kilometres per second and the end of the Cretaceous was three hours away. It seemed to hang in the sky like a second moon, or the eye of God, but no dinosaur looked at it with understanding. It suddenly swelled in the sky, and then a dark mantle spread across the firmament.

Why did it have to be this way?

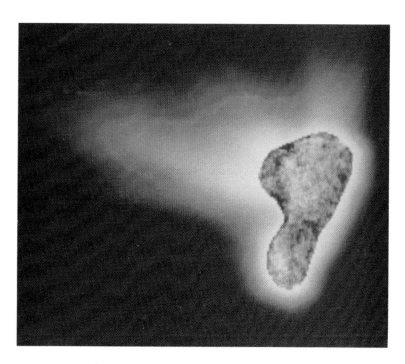

Halley's Comet:
peanut-shaped nucleus (8 by 15 kilometres),
photographed by Soviet *Vega* spacecraft

Earthrise over the Moon

A planetary substrate,
suitable for complex organisms,
as seen from space

The Meaning of the Dinosaurs

Life on a Broader Scale

Perhaps the extinction of the dinosaurs really didn't matter.

To understand why this may be so, it is useful to review briefly the circumstances under which dinosaurian evolution took place. Dinosaurs lived on a single planet, a tiny fleck in a cosmos that is vast beyond our comprehension. As we do now, dinosaurs once had the great gift of being alive, but they lived in a world that differed biologically in interesting ways from our own. Of their world, only traces remain in sedimentary record. We can attempt to appreciate these differences by examining limited information and using the only time machine we possess – a limited imagination. The importance of the dinosaurian world to us is rooted in the relationship of life to the material universe and in the presence of pattern in the way life changes with time.

The observable universe is both vast and homogeneous, and the same physical and chemical laws apply everywhere. If the relationships between the basic components of matter differed even slightly from what they are, life would not exist. A fine balance between nuclear forces permitted hydrogen to become the most abundant element, and thereby fuel the stars. The ratio of light particles to nuclear particles was low enough to allow stars to condense. Gravity was strong enough to force stars to generate carbon and oxygen in nuclear reactions in their cores, and then explode, seeding space with the elements on which organic chemistry is based. It was weak enough so that protoplanets would not fall into their stars and could exist as separate objects. The balance between gravity and the expansion of the universe has allowed it to attain an age three times that of the Sun, so that sufficient time for complex life to evolve exists in abundance. The universe is fundamentally compatible with life.

Many organic compounds have been identified by radioastronomical techniques in interstellar space. These complex molecules can escape the high temperatures of star and planet formation within the gigantic halo of cold protocometary bodies left behind around new planetary systems. Key chemical building-blocks in the genetic code have been identified in meteorites, which have fallen to Earth following the disintegration of comets and asteroids in nearby space. Life was present on Earth a few hundred million years after the surface cooled, nearly four thousand million years ago. The implications are that life rapidly originates on Earth-like planets and that chemical processes leading to life are operative everywhere. The universe appears to be constructed so that life will appear within it, although the processes by which it originates remain obscure.

For thousands of millions of years the surface of the Earth has provided benign environments in which simple cells became more complex, complex cells formed colonies, and simple organisms became ever more complicated. Is it the only place where organic evolution has occurred? Dark material has been detected in orbit around nearby stars, and stars often occur in small groups that orbit a common centre of gravity. These circumstances suggest that planets are not rare beyond our solar system, in spite of the fact that they have not yet been seen. Most of the more than 100 billion stars in the Milky Way Galaxy could not sustain life on Earth-like planets. Large hot stars burn out too quickly, and small cool ones can warm only planets that are very near. The rotation of these planets is soon stopped, and their atmospheres are lost. Among the most suitable stars are the approximately 20 billion Sun-like stars in the Milky Way Galaxy. Many of these rather average stars belong to double- or multiple-star systems, where planetary orbits are often unstable. However, about one in ten is solitary, like the Sun. If one planet in forty occurs in a proper orbit

Venus, an Earth-like planet
in an orbit slightly too close
to the Sun to support
complex organisms

Mars: probably too distant from Sun
to support complex organisms.
The occurrence of three Earth-like bodies in one
system implies that such planets may not be rare.

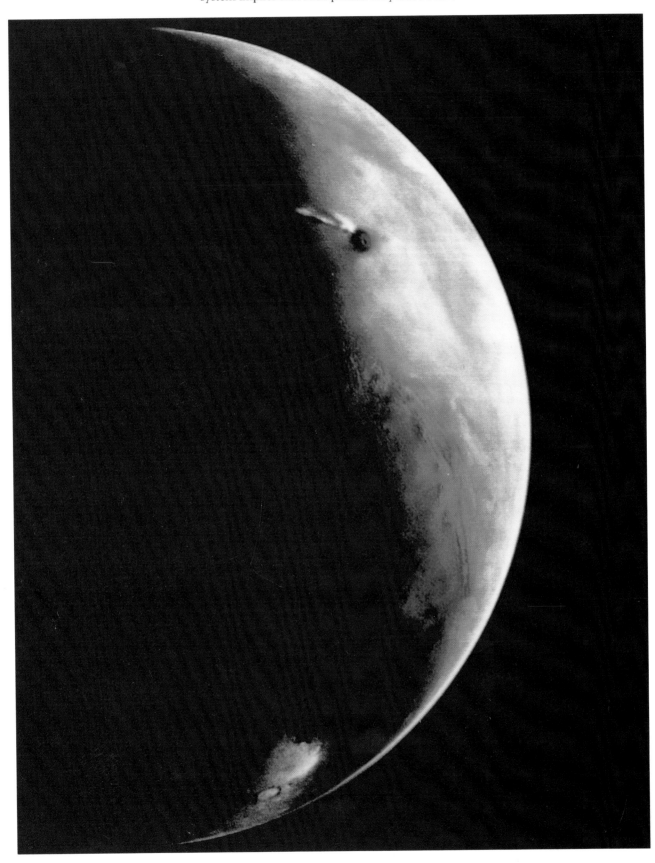

around a Sun-like star, there would be 50 million Earth-like planets in our galaxy. There are an incredible number of galaxies in the cosmos. The universe would appear to be permeated with life.

Life on Earth

Over 500 million years ago, a jellyfish was a relatively complicated organism, but now a flying fish is a relatively complicated organism. In a comparison between a jellyfish and a flying fish, progress is obvious in the increased behavioural complexity of the latter, which is made possible by a greater number of harmoniously functioning parts. If there exists a general pattern of organic change through time from the simple to the more complex, it follows that some kind of order is present in the history of life. The presence of evolutionary order can be expected from the discovery of natural order in all branches of science.

Evolutionary change is seen usually as being the result of competition between organisms. In order to survive and reproduce, an organism must overcome challenges both from its physical and from its biological surroundings. For example, European house sparrows were introduced to North America a little over one hundred years ago and have divided into larger northern races (subspecies) and smaller southern ones. A larger body size enables the northern races to resist low temperatures but this does not imply evolutionary progress (in the sense that the flying fish is more complicated than the jellyfish). Conversely, carnivorous mammals living in forests have longer limbs, more slender bodies, and larger brains than did their predecessors living in similar environments 30 million years ago. During the same interval the physical environment changed little. The modern carnivores, as a result of biological interactions, have made evolutionary progress.

The importance of the biological environment in evolution is apparent on a larger scale. During the past billion years the numbers of different species living on our planet have increased at an ever more rapid rate. This increase implies that communities of organisms have also grown in complexity. Yet, from an astronomical point of view, physical environments on the surface of the Earth have changed very little. Because the biosphere has become much more complex, the increase in complexity was more probably driven by interactions between different organisms than those between the organisms and their physical environment.

During the past 500 million years on Earth, biological complexity did not increase at a constant rate. Rather, it grew at an ever-quickening pace, apparently proportional to the level of complexity already attained. The process is similar to one described in a time-honoured story of a man who asked an employer to agree to pay him one cent for the first day's work, but to continue to double his salary on each subsequent day for one month. Thus his rate of pay would be proportional to (in this case double) the rate he was paid on the previous day. On the tenth day his salary would rise to $5.12, on the twentieth to $5,242.88, and on the thirtieth to $5,368,709.12. If the employer had bargained for an increase of 25 per cent per day, then he would have had to pay only $6.46 on the thirtieth day. In both cases, however, the rate of increase would have been proportional to the level attained. Just as the pay rate was driven by daily salary level received, so too the pace of evolution seems to be driven by evolutionary level attained. The intensity of interaction between organisms and their biological environment thus sets the pace of evolution.

Examples of evolutionary changes that occurred at ever-increasing speeds include the initial diversification of animals in the sea between 650 and 550 million years ago, the attainment of tree stature in land plants between 410 and 360 million years ago, and the diversification of mammals between 200 million years ago and the present. A similar increase in weights attained by organisms continued until about 100 million years ago. Changes like these have resulted in increased organismal complexity, which, in combination with a general increase in number of species, has made the biosphere of the modern Earth so much richer than it was several hundred million years ago. It is reasonable to suppose that animals living in a complex environment might find it advantageous to possess complex nervous systems in order to have access to a greater variety of responses. Indeed, the largest proportion of body weight devoted to brain tissue in an animal has also increased at an ever-increasing rate through geological time. The brain has evidently become larger in animals as diverse as insects, molluscs, and backboned creatures. Relative brain size can be taken as an indication of the complexity of biotic interactions.

Organisms living in similar environments and interacting with other organisms in similar ways tend to become similar through geological time. This is because natural selection favours the survival of organisms that more fully possess qualities demanded by an ecological niche. These qualities describe an efficient form that in turn provides a target for natural selection. Birds, bats, and flying fishes resemble each other because of aerodynamic qualities required for flight. In a process that has been called convergent evolution, creatures of different origin have acquired body forms that mimic common ancestry. For example, in Australia several tens of millions of years ago, primitive crows evolved into mimics of warblers, flycatchers, and nuthatches. The resemblance is so close that these birds were often classified with their European counterparts until molecular-genetic analyses demonstrated their closer relationship to crows. Similarities in the appearance and anatomy of pandas were considered as evidence linking the giant panda to the racoon family through the lesser panda. Molecular-genetic studies, however, demonstrated that giant pandas are related more closely to bears.

These are extreme examples of convergence, which are heightened by the fact that the animals involved were originally rather closely related. Convergence is more obvious in more distantly related organisms, such as Canadian wolves and Tasmanian marsupial wolves, Amazonian capybaras and west African chevrotains, and South American armadillos and Asian pangolins. The effects of convergence, however, can be seen in very distantly related organisms. Squids and herrings are derived from utterly different body plans. One swims by lateral undulations of the body with the head in front, and the other swims with a water jet, with its head to the rear. Yet these swimming creatures have come to resemble each other increasingly through geological time. Their rapid movements, fusiform shapes, large eyes, coloration, and gregarious habits make it difficult to distinguish schools of these creatures in the surface waters of the open ocean.

The Evolution of Dinosaurs

Dinosaurs dominated the lands of our planet for nearly 150 million years. Their skeletal parts became more differentiated through time, and their activity levels and behavioural complexity probably also increased. However, their evolutionary accomplishments must be understood in terms of the biological environment of their time, not that of the modern era. Using the brain guide to biospheric complexity, evolutionary progress achieved during the dinosaurian era would have been accomplished within 30 million years under evolutionary pressures generated by a modern biosphere.

The Triassic world was as dominated by quadrupeds as is our own, and bipedal dinosaurs would have appeared about as out of place in it as we do in a modern African game reserve. At the close of the Triassic, most of the dinosaurs' large quadrupedal contemporaries vanished in a mysterious extinction which they were able to survive. Because bipedal locomotion is energetically as efficient as quadrupedal locomotion, there was no unusually strong pressure for dinosaurs to become quadrupedal. The dinosaurian era was accordingly a time of bipedality, and as a consequence dinosaurian skeletal parts often seem odd by modern mammalian standards. Suppose that birds were to become terrestrial quadrupeds and readapt their wings for use as forelimbs. Analogous problems were overcome by some dinosaurs in their adoption of quadrupedality. In spite of an embryology that was fundamentally that of a biped, however, many adult dinosaurs were constructed in ways that recall the forms of some modern mammals. There were also convergences with modern birds.

The resemblance is obvious between horned dinosaurs and the mammalian rhinos that later imitated them, as well as between armoured ankylosaurs with tail clubs and some of the recently extinct giant armadillos which also possessed tail clubs. It is difficult to watch a dromedary without visualizing the form of a hadrosaur, in spite of the mammal's quadrupedal stance and vestigial tail. The dromaeosaurs must have resembled flightless eagles that bore claws on their featherless arms. Their temperaments were probably also similar. Toothless oviraptors had the heads of parrots, and their brains were relatively large as well, by dinosaurian standards. Tyrannosaurs bore the heads and necks of wolves on bodies vaguely resembling those of ground birds. There was an even greater resemblance between ornithomimids and ornithomimid-mimics, the ostriches. Recall the comparison between a sauropod dinosaur and an elephant. Some anatomical interposition was required, because the mammal uses teeth to grind its food, whereas the dinosaur used stomach stones. The elephant's head is so heavy that the neck must remain short in order to provide adequate support. It accordingly possesses a very long nose which duplicates the function of the dinosaur's long neck. The evolutionary convergences between dinosaurs and modern mammals and birds were at least as great as those between fishes and squids.

Because their extinction was probably a result of the impact of a cometary nucleus or an asteroid, the dinosaurs were overwhelmed by a bizarre set of physical stresses that were very unlike the relatively stable physical and biological stresses under which their evolution had taken place. The stresses were non-biological in origin and produced a colossal dying in which many different species became totally extinct. Instead of leading to greater complexity, the extinctions produced a simplification of ecosystems. Many millions of years would pass before increasing anatomical sophistication in surviving organisms brought terrestrial ecosystems to their former levels of complexity. Had another extinction occurred during the period of recovery, the resulting simplification might again have brought complexity levels back to the point where they were after the previous extinction. Mass extinctions are not evolutionarily 'positive' events.

Although the extinction did not result in progress, it did produce evolutionary 'noise.' Without the impact of the cometary body, dinosaurs would not have become extinct. It did strike, and as a consequence quadrupedal mammals gradually occupied the ecological roles vacated by the dinosaurs. What was a disaster for the terrestrial biosphere was a boon for mammals. Their diversification took place in an impoverished, unbalanced ecological setting where the physical environment loomed large. Surviving mammals responded like the European house sparrows in North America a century ago, by separating into races according to local surroundings. They could not immediately compensate for the loss in anatomical richness produced by the long-established biological interactions from which the dinosaurs had benefited. But the great drama of the evolution of life on land did continue, with one group of players (dinosaurs) suddenly being replaced by a second string (mammals). We are the descendants of the second string.

Similarities had occurred between the anatomy of unrelated dinosaurs living on different continents, such as between the Cretaceous Northern Hemisphere dromaeosaurs and South American dromaeosaur mimics, the noasaurs. Similarities in body form can be seen today between unrelated mammals that evolved on different continents, and these were responsible for originally attracting attention to the phenomenon of evolutionary convergence. Convergences confirm the presence of pattern in evolution, for in each case the form in question represents a target favoured by natural selection in the struggle for existence. Because the dinosaurs were not eliminated as a consequence of the struggle for existence, but by a tremendous, extraterrestrially induced catastrophe, they can be looked on as another great biohistorical evolutionary experiment, one more grand even than the separate evolution of mammals on different continents.

Comparisons between dinosaurs and mammals thus provide another means of evaluating convergences, or evolutionary 'signal.' Conversely, the effects of their peculiar bipedal origins, as opposed to mammalian quadrupedalism, offer a means of identifying divergences from patterns, or evolutionary 'noise.' The balance between evolutionary 'signal' and evolutionary 'noise' in the dinosaurian and mammalian evolutionary experiments may provide a guide to what might be expected to occur in the evolution of land-dwelling multicellular creatures anywhere in the cosmos.

A Thought Experiment

The dinosaurs were evidently unable to survive the intense but short-term climatic stresses caused by the impact of a cometary body. They were not, however, failures in the biological struggle for existence, and their evolutionary history at least in part presaged that of the mammals that subsequently replaced them. An obvious question would be: what would have happened if the cometary body had not struck the Earth and the dinosaurs were not exterminated? It cannot be answered with certainty, because our knowledge is so incomplete. But regularities in the history of life invite a mental exercise of the kind that has been called a 'thought experiment.' Such an 'experiment' might not be accurate, but others have been useful in that they have posed questions that subsequently could be answered.

The biosphere of our planet has not subsequently suffered a catastrophe as profound as the one that took place at the end of the age of reptiles. It can therefore be presumed that dinosaurs would still be the dominant animals on land. With increasing metabolic rates, their size would have continued to diminish in order to balance bodily heat budgets. Under the pruning influence of natural selection, many would have come to resemble many mammals more closely in basic body form. The mammals themselves, however, would have continued to occupy the insectivore and rodent niches as they indeed did throughout the Mesozoic Era.

One very obvious trend in the evolution of mammals since the end of the dinosaurian age was a threefold increase in average brain-body proportions. The scatter in brain proportions also increased, so that in some living mammals (opossums) they are as small as in Mesozoic mammals, while in others (humans) they are nearly thirty times larger. At the time of their extinction, the brain-body proportions in some dinosaurs were equal to those in their mammalian contemporaries. Is it not reasonable to suppose that mean relative brain size and the scatter in relative brain size would have similarly increased in dinosaurs during the 65 million years since the end of the Cretaceous?

During the latter part of Cretaceous time, many different varieties of small carnivorous dinosaurs possessed brains that were quite large relative to those of living reptiles. One of these, *Troodon*, was rather unspecialized and can be taken

Model of small theropod dinosaur *Troodon*

as typical of these small theropods. Incorporated into its skeleton were characteristics in which it resembled monkeys more than did the Cretaceous mammals that were actually in the ancestry of primates, including bipedality, a flexible and opposable digit in the hand, and huge eyes with overlapping visual fields. It is biologically possible for creatures with human brain-body proportions to exist, because they (we) do exist. For the purposes of the thought experiment, what might some anatomical consequences be for the existence of such brain-body proportions within the body plan of a small theropod dinosaur? They would not be arbitrary, for it they possessed no functional purpose they would not be targeted by selective pressures.

The anatomy of a hypothetical large-brained dinosaur could be divided into three components. One set would be associated with the possession of a large brain, whether it occurred in a mammal, a bird, or a dinosaur. Another set of characteristics would stem from the fact that the hypothetical large-brained creature would have had a dinosau-

rian (*Troodon*) ancestry. A third set of characteristics might be more general ones associated with a reptilian as opposed to a mammalian body plan.

One consequence of a large brain is that the rest of the skull would seem relatively small. This is obvious in embryonic lizards, which provide a model for braincase-facial proportions in a large-brained reptile. The head will also be large in relation to the body, and this will be particularly obvious in animals in the 30-to-70-kilogram range. A large head is supported most easily by a vertical neck centred beneath it. The adaptation is apparent in apes, and also in Cretaceous pachycephalosaurs, which carried small brains in heavy, bony skulls. A vertical backbone provides an energetically efficient posture for carrying the upper part of the body and a heavy head, although a horizontal backbone provides a better framework for strong running muscles. Ostriches are bipedal, like the small theropod dinosaurs of long ago, and carry their backbones in a horizontal position. The large head on an ostrich would be tiresome to carry

Head of hypothetical,
large-brained descendant of *Troodon*

and exposed to injury. The human leg has been the target of strong selective forces and is an extremely efficient structure for walking. It does not, however, enable humans to run as rapidly as most large mammals do, and our running abilities would not ensure our survival in an African game reserve. Instead, the brain has been substituted for an ability to run rapidly as a first-line means of self-preservation. It is recommended that one use one's brain in an African game reserve.

The pelvic canal is very large in some small theropods, and they may have given birth to living young. If the young of a hypothetical large-brained dinosaur were carried in an egg, the embryonic support tissues originally needed to support the growing embryo would necessarily be heavier than the hatchling alone. The pelvic canal would in turn have to be wider in order to allow a relatively larger egg to pass. Conversely, if the young were born alive the hips could be narrower, and the efficiency of walking thereby enhanced. Many living lizards and snakes give birth to living young,

and our creature may be considered as having once been nurtured through an umbilical cord. It is also somewhat daringly provided with a navel. The brain is an energetically demanding organ, requiring one-fifth of our total resting metabolism. A human brain would drain a crocodile to the point where the animal would have no metabolic energy left over for moving. Our hypothetical creature would accordingly have a high metabolic rate and be warm to touch. It would then have to eat often to maintain its energy balance. Here a separation of the mouth and nasal passages would be useful so that it could eat and breathe at the same time. This 'secondary palate' is present in modern mammals and was present in ostrich dinosaurs many millions of years ago.

The effect of a very large brain in a small dinosaur might thus be in the direction of convergence with large-brained mammalian bipeds. However, genetic relationships are revealed usually by similarities in anatomical detail, and this correlation implies strongly that the effects of a theropod ancestry would be very obvious. The teeth were reduced in *Troodon* and its relatives, and in at least two groups of small theropods they were lost entirely. Teeth are absent in modern birds, which were possibly descended from small theropod dinosaurs. The jaws of the large-brained dinosaur are thus postulated to be toothless and provided with keratinous biting and chewing surfaces. The proportions of its arms are taken from those of ostrich dinosaurs. As in most small theropods, there are only three fingers in the hand. Theropods did not have kneecaps, and the muscles of the upper leg are accordingly attached to a blade on the upper part of the shin-bone. The foot is a compromise between the foot in a small theropod and the foot in a tree kangaroo, which are similar in that the small toes are on the inside, not the outside of the foot.

Other anatomical characteristics of our hypothetical large-brained theropod would be those more typical of dinosaurs or more distantly related reptiles. These might include large eyes, the absence of ears, the presence of a dewlap as a secondary sexual characteristic, the nourishment of the young not with milk but by providing regurgitated, partially digested food, and the containment of external genitalia within a pelvic pouch.

In general appearance, a large-brained dinosaur may have resembled a large-brained bipedal mammal. The convergence in body form might well have been greater than that between pelagic squids and fishes. Should a high degree of convergence have been attained, the adaptive meaning of differences between the dinosaur and mammal would be minimal. From a biological point of view, they would be equivalent alternatives. And the extinction of the dinosaurs would really not have mattered.

How to Live in Space-Time

In the course of our lifetimes we will not be able to study distant biospheres beyond the solar system. Are there quasi-dinosaurs living in ecosystems of far-off planets? If convergence is an important phenomenon in the history of life, and if the dinosaurian world was a distinct experiment in the evolution of complex life on Earth, then dinosaur-like 'noise' has a certain probability of appearing in exotic biospheres. This probability may not be high, for the ascendancy of the dinosaurs on Earth may have depended on the extinction (non-competitive removal) of the ancient Triassic quadrupeds. Without the extinctions at the end of the Triassic, the transition from a reptile-dominated to a mammal-dominated world would have been much smoother, and dinosaurs would have been just a peculiar side-branch of bipedal crocodiloids that gave rise to birds. The dinosaurs, as we know them, exhibit many anatomical details that link them to the unique expression of the development of life on Earth. Like *Homo sapiens, Tyrannosaurus rex* carried within it the stamp of terrestrial origin. Extraterrestrial 'dinosaurs' would not carry this stamp.

Our life-spans are much too limited to follow the progress of life on our own planet into a distant future. It is conceivable that organic complexity on the present Earth can never be surpassed, but there is no firm evidence that this is so. The long-term transit of dinosaurs and other living things to higher levels of complexity may contain other trends that can be extrapolated in order to discern the ever more complex terrestrial biospheres to come. If the past can be a guide, evolutionary rates of change will increase, so that a given level of change in biological complexity will require less and less time to be achieved. While improvements will continue in physiology and anatomical structure, neurophysiological selection will become increasingly important.

There will be failures. In the past, southern man-apes (australopithecines) became extinct, Neanderthal peoples disappeared, but modern humans are now tremendously successful organisms. If we should become extinct, will not natural selection drive surviving primates, racoons, rats, or perhaps even parrots toward higher levels of manipulative and neurological ability? Perhaps the atmosphere of our planet will not be able to compensate for the gradually increasing brilliance of the Sun, so that our planet will become uninhabitable after another 100 million years. But within the last few decades technologies have been invented that allow the survival of humans in space for at least limited periods of time. Will technological progress cease, so that for 100 million years to come human populations will never be able to colonize extraterrestrial environments?

There will be surprises that are as difficult to foresee as analogous evolutionary advances might have been in the past, which nevertheless have already taken place. Eons ago carbon and hydrogen atoms combined to form the genetic code and made inheritance possible. Primitive cells fused together to form advanced, nucleated cells of the kind that make up multicellular organisms. Fungi merged with marine algae to form plants that could live on land. How could it have been predicted 500 million years ago that jellyfish-like animals would give rise to animals with brains? Or, from the fossil record, that the jelly-like brain of proto-man would produce a technological civilization?

Models of *Troodon* (RIGHT)
and a hypothetical,
large-brained descendant

A Perspective That Includes Dinosaurs

Dinosaurs are interesting because of their large size and unfamiliar appearance. They stimulate our imaginations and help us to consider some rather basic questions about the world in which we live. Was the evolution of dinosaurs on Earth an average, more or less typical process? Did dinosaurs assume more or less predictable shapes because of the effects of convergent selection? Should these apparently preferred evolutionary pathways be more carefully studied, in order to help us understand the evolution of complex life in space and time? Or was the dinosaurian world somehow special? Could it better serve to help us understand how much variation exists around preferred evolutionary pathways? Is the form of a hypothetical large-brained dinosaur so astonishing from an evolutionary point of view? There is much discussion today about the possibility of communicating with extraterrestrial intelligence through signals transmitted by radiotelescopes. Might a biologist advise his radio-astronomer friend not to be too amazed if an organism returning a message from space has a rather familiar look?

A century ago we did not know very much about dinosaurs. Now we know more. There was really no reason, a century ago, for specialists to be afraid of what has subsequently been discovered. Nor is there any reason for us to be anxious about what will be learned in the century to come. If what we do not yet understand is like what we do understand about life, then the unknown will be challenging, interesting, and beautiful.

Bibliography

The references cited below are arranged according to the chapters or groups of chapters in which data they contain have been cited. They are a necessary minimum needed to support the text, and each reference contains further lists of citations that could usefully be consulted. References preceded by an asterisk (*) present points of view in disagreement with those expressed in the text.

Preface

Barron, E.J. (ed.). 1987. Special Issue, Cretaceous Paleogeography. *Palaeogeography, Palaeoclimatology, Palaeoecology*, volume 59. 214 pp.

Bird, R.T. 1985. *Bones for Barnum Brown.* Texas Christian University Press, Fort Worth. 225 pp.

Calder, W.A. 1984. *Size, Function and Life History.* Harvard University Press, Cambridge, Mass. 431 pp.

Charig, A., and B. Horsfield. 1975. *Before the Ark.* British Broadcasting Corporation, London. 160 pp.

Colbert, E.H. 1968. *Men and Dinosaurs.* Dutton, New York. 283 pp.

Desmond, A.J. 1975. *The Hot-Blooded Dinosaurs: A Revolution in Palaeontology.* Blond and Briggs, London. 238 pp.

Hammond, J.K., and P. Dodson. 1986. 'Those Who Study Dinosaurs Face Extinction.' *New York Times*, 26 November.

Harland, W.B., A.V. Cox, P.G. Llewellyn, C.A.G. Pickton, A.G. Smith, and R. Walters. 1982. *A Geologic Time Scale.* Cambridge University Press, London. 131 pp.

McGowan, C. 1983. *The Successful Dragons.* Samuel Stevens, Toronto. 263 pp.

Norman, D. 1985. *The Illustrated Encyclopedia of Dinosaurs.* Crescent Books, New York. 208 pp.

Owen, H.G. 1983. *Atlas of Continental Displacement, 200 Million Years to the Present.* Cambridge University Press, Cambridge. 159 pp.

Russell, D.A. 1987. 'Models, Paintings and the Dinosaurs of North America.' In Sylvia J. Czerkas and Everett C. Olson (eds.), *Dinosaurs Past and Present*, Vol. 1, pp. 115–31. Natural History Museum of Los Angeles County/University of Washington Press, Seattle and London.

Simpson, G.G. 1942. 'The Beginnings of Vertebrate Paleontology in North America.' *Proceedings of the American Philosophical Society* 86: 130–88.

Before the Mesozoic

(CHAPTER ONE)

Anderson, J.M., and A.R.I. Cruickshank. 1978. 'The Biostratigraphy of the Permian and the Triassic. Part 5. A Review of the Classification and Distribution of Permo-Triassic Tetrapods.' *Palaeontologia Africana* 21: 15–44.

Carroll, R.L. 1976. 'Eosuchians and the Origin of Archosaurs.' In C.S. Churcher (ed.), *Athlon: Essays on Palaeontology in Honour of Loris Shano Russell*, pp. 58–79. Royal Ontario Museum Life Science Series Miscellaneous Publication.

~ 1982. 'Early Evolution of Reptiles.' *Annual Review of Ecology and Systematics* 13: 87–109.

Carroll, R.L., E.S. Belt, D.L. Dineley, D. Baird, and D.C. McGregor. 1972. *Excursion A59, Vertebrate Paleontology of Eastern Canada.* 24th International Geological Congress, Montreal. 112 pp.

Raup, D.M., and J.J. Sepkoski Jr. 1982. 'Mass Extinctions in the Marine Fossil Record.' *Science* 215: 1501–3.

Schopf, J.W. (ed.). 1983. *Earth's Earliest Biosphere: Its Origin and Evolution.* Princeton University Press, Princeton. 543 pp.

The Triassic

(CHAPTER TWO)

Ash, S.R. 1978. 'Geology, Paleontology, and Paleoecology of a Late Triassic Lake, Western New Mexico.' *Brigham Young University, Geological Studies* 25 (2): 1–100.

~ 1980. 'Upper Triassic Floral Zones of North America.' In D.L. Dilcher and T.N. Taylor (eds.), *Biostratigraphy of Fossil Plants*, pp. 153–70. Dowden, Hutchinson and Ross, Stroudsburg, Pa.

~ 1985. 'A Short Thick Cycad Stem from the Upper Triassic of Petrified Forest National Park, Arizona, and Vicinity.' *Museum of Northern Arizona Bulletin* 54: 17–32.

~ 1986. *Petrified Forest: The Story behind the Scenery.* Petrified Forest Museum Association, Arizona. 48 pp.

Baird, D., and P.E. Olsen. 1983. 'Late Triassic Herpetofauna from the Wolfeville Fm. of the Minas Basin (Fundy Basin) Nova Scotia, Can.' Geological Society of America, 18th Annual Meeting, Northeastern Section, Abstracts with Program 15: 122.

Baird, D., and O.F. Patterson. 1967. 'Dicynodont-Archosaur Fauna in the Pekin Formation (Upper Triassic) of North Carolina.' Geological Society of America Meeting in New Orleans, Abstracts, p. 11.

Bakker, R.T., and P.M. Galton. 1974.

'Dinosaur Monophyly and a New Class of Vertebrates.' *Nature* 248: 168–72.

Balkwill, H.R., D.G. Cook, R.L. Detterman, A.F. Embry, E. Hakansson, A.D. Miall, T.P. Poulton, and F.G. Young. 1983. 'Arctic North America and Northern Greenland.' In M. Moullade and A.E. Nairn (eds.), *The Phanerozoic Geology of the World*, II *The Mesozoic*, pp. 1–31. B. Elsevier Science Publishers, Amsterdam.

Benton, M.J. 1983. 'Dinosaur Success in the Triassic: A Noncompetitive Ecological Model.' *Quarterly Review of Biology* 58: 29–55.

~ 1984. 'Consensus on Archosaurs.' *Nature* 312: 599.

~ 1984. 'Rauisuchians and the Success of Dinosaurs.' *Nature* 310: 101.

~ 1985. 'Classification and Phylogeny of the Diapsid Reptiles.' *Zoological Journal of the Linnean Society* 84: 97–164.

Benton, M.J., and A.D. Walker. 1985. 'Palaeoecology, Taphonomy, and Dating of Permo-Triassic Reptiles from Elgin, Northeast Scotland. *Palaeontology* 28: 207–34.

Camp, C.L. 1930. *A Study of the Phytosaurs.* University of California Memoirs 10. 161 pp.

~ 1980. 'Large Ichthyosaurs from the Upper Triassic of Nevada.' *Palaeontographica* Abt. B, Band 170: 139–200.

~ 1981. *Child of the Rocks.* Nevada Bureau of Mines and Geology, Special Publication 5. 35 pp.

Camp, C.L., and S.P. Welles. 1956. 'Triassic Dicynodont Reptiles.' *University of California Memoirs* 13: 225–348.

Casamiquela, R.M. 1980. 'La presencia del genero *Plateosaurus* (Prosauropoda) en el Triasico Superior de la Formacion el Tranquilo, Patagonia.' I *Congreso Latinamericano de Paleontologia, Buenos Aires*: 143–58.

Charig, A.J. 1984. 'Competition between Therapsids and Archosaurs during the Triassic Period: A Review and Synthesis of Current Theories.' *Symposia of the Zoological Society of London* 52: 597–628.

Charig, A.J., B. Krebs, H.D. Sues, and F. Westphal. 1976. 'Thecodontia.' *Handbuch der Palaoherpetologie* 13: 1–137.

Chatterjee, S. 1984. 'A New Ornithischian Dinosaur from the Triassic of North America.' *Naturwissenschaften* 71: 630–1.

~ 1985. '*Postosuchus*, a New Thecodontid Reptile from the Triassic of Texas and the Origin of Tyrannosaurs.' *Philosophical Transactions of the Royal Society of London* B309: 395–460.

~ 1986. '*Malerisaurus langstoni*, a New Diapsid Reptile from the Triassic of Texas.' *Journal of Vertebrate Paleontology* 6: 297–312.

~ 1986. 'News ("Fossil Bird Shakes

Evolutionary Hypotheses").' *Nature* 322: 677.

Colbert, E.H. 1946. '*Hypsognathus*, a Triassic Reptile from New Jersey.' *Bulletin of the American Museum of Natural History* 86: 225–74.

~ 1957. 'Triassic Vertebrates of the Wind River Basin.' *Wyoming Geological Association Guidebook, 12th Annual Field Conference*, pp. 89–93.

~ 1970. 'The Triassic Gliding Reptile *Icarosaurus*.' *Bulletin of the American Museum of Natural History* 143: 85–142.

~ 1974. 'The Triassic Paleontology of Ghost Ranch.' *New Mexico Geological Society Guidebook, 25th Field Conference*, pp. 175–8.

~ 1980. *A Fossil-hunter's Notebook*. E.P. Dutton, New York. 242 pp.

~ 1982. 'The Distribution of *Lystrosaurus* in Pangea and Its Implications.' *Geobios, mémoire special* 6: 375–83.

~ 1982. 'Triassic Vertebrates in the Transantarctic Mountains.' *Antartic Research Series* 36: 11–35.

~ 1985. 'The Petrified Forest and Its Vertebrate Fauna in Triassic Pangea.' *Museum of Northern Arizona Bulletin* 54: 33–43.

Colbert, E.H., and J. Imbrie. 1956. 'Triassic Metoposaurid Amphibians.' *Bulletin of the American Museum of Natural History* 110: 399–452.

Crowley, T.J. 1983. 'The Geologic Record of Climatic Change.' *Reviews of Geophysics and Space Physics* 21: 828–77.

Cruickshank, A.R.I., and M.J. Benton. 1985. 'Archosaur Ankles and the Relationships of the Thecodontian and Dinosaurian Reptiles.' *Nature* 317: 715–17.

Dawley, R.M., J.M. Zawiskie, and J.W. Cosgriff. 1979. 'A Rauisuchid Thecodont from the Upper Triassic Popo Agie Formation of Wyoming.' *Journal of Paleontology* 53: 1428–32.

Dedera, D. 1983. 'Arizona's Other National Park.' *Arizona Highways* 59 (2): 2–34.

Delevoryas, T. 1970. 'Plant Life in the Triassic of North Carolina.' *Discovery* 6: 15–22.

Dickinson, W.R. 1980. 'Cretaceous Sinistral Strike Slip along Nacimiento Fault in Coastal California.' *American Association of Petroleum Geologists Bulletin* 67: 624–45.

Galton, P.M. 1977. 'On *Staurikosaurus pricei*, an Early Saurischian Dinosaur from the Triassic of Brazil, with Notes on the Herrerasauridae and Poposauridae.' *Palaontologische Zeitschrift* 51: 234–45.

~ 1983. 'The Oldest Ornithischian Dinosaurs in North America from the late Triassic of Nova Scotia, N.C. and PA.' 18th Annual Meeting, Geological Society of America, Northeastern Section, Abstracts with Programs, p. 122.

Gottesfeld, A.S. 1972. 'Paleoecology of the

Lower Part of the Chinle Formation in the Petrified Forest.' *Museum of Northern Arizona Bulletin* 47: 59–73.

Gregory, J.T. 1969. 'Evolution und interkontinentale Beziehungen der Phytosauria (Reptilia).' *Palaontologische Zeitschrift* 43: 37–51.

Hopson, J.A. 1984. 'Late Triassic Traversodont Cynodonts from Nova Scotia and Southern Africa.' *Palaeontologia Africana* 25: 181–201.

Hubert, J.F., and K.A. Mertz. 1980. 'Eolian Dune Field of Late Triassic Age, Fundy Basin, Nova Scotia.' *Geology* 8: 516–19. (See discussion in *Geology* 9: 557–9.)

Hughes, D.W. 1981. 'Why Is the Moon Slowing down?' *Nature* 290: 190.

Hyde, M.G. 1981. 'Stratigraphy, Petrology, and Depositional Environment of the Upper Triassic Blomidon Formation, St. Mary's Bay, Nova Scotia.' MSC thesis, Department of Geology and Geography, University of Massachusetts, Amherst. 240 pp.

Jacobs, L.L. (ed.). 1980. *Aspects of Vertebrate History*. Museum of Northern Arizona Press, Flagstaff. 407 pp. (See articles by L.L. Jacobs and P.A. Murray and by P.E. Olsen.)

Jacobsen, V.W., and P. van Veen. 1984. 'The Triassic Offshore Norway North of 62° N.' In A.M. Spencer et al (eds.), *Petroleum Geology of the North European Margin*, pp. 317–27. Graham and Trotman, Trondheim Norway.

Kahn, P.G.K., and S.M. Pompea. 1978. 'Nautiloid Growth Rhythms and Dynamical Evolution of the Earth-Moon System.' *Nature* 275: 606–11. (See discussion in *Nature* 279: 452–6.)

Kues, B.S., K.K. Kietzke, and S.G. Lucas. 1984. 'A New Late Triassic Invertebrate and Vertebrate Fauna from the Chinle Formation, Eastern New Mexico.' Geological Society of America, Abstracts with Program 16 (4): 277.

Long, R.A., and K.L. Ballew. 1985. 'Aetosaur Dermal Armor from the Late Triassic of Southwestern North America, with Special Reference to Material from the Chinle Formation of Petrified Forest National Park.' *Museum of Northern Arizona Bulletin* 54: 45–68.

Love, J.D. 1957. 'Stratigraphy and Correlation of Triassic Rocks in Central Wyoming.' *Wyoming Geological Association, 12th Annual Field Conference*, pp. 39–45.

Lucas, S.G., K.K. Kietzke, J. Sobus, and G. Weadock. 1984. 'Upper Triassic–Upper Jurassic Stratigraphy, Fossil Vertebrates and Depositional Environments, Bull Canyon, Guadalupe County, New Mexico.' Geological Society of America, Abstracts with Program 16 (4): 245.

Manspeizer, W., J.H. Puffer, and H.L. Cousminer. 1978. 'Separation of Morocco and Eastern North America: A Triassic-Liassic Stratigraphic Record.' *Geological Society of America Bulletin* 89: 901–20.

McGowen, J.H., G.E. Granata, and S.J. Seni. 1979. *Depositional Framework of the Lower Dockum Group (Triassic), Texas Panhandle*. Texas Bureau of Economic Geology, Report of Investigations 97. 60 pp.

McKee, E.D., S.S. Oriel, K.B. Ketner, M.E. MacLachlan, J.W. Goldsmith, J.C. MacLachlan, and M.R. Mudge. 1959. *Paleotectonic Maps of the Triassic System*. U.S. Geological Survey, Miscellaneous Geologic Investigations Map I-300. 33 pp.

Meyer, L.L. 1986. 'D-Day on the Painted Desert.' *Arizona Highways* 62 (7): 2–13.

Olsen, P.E. 1986. 'A 40 Million Year Lake Record of Early Mesozoic Orbital Climatic Forcing.' *Science* 234: 842–8.

Olsen, P.E., A.R. McCune, and K.S. Thomson. 1982. 'Correlation of the Early Mesozoic Newark Supergroup by Vertebrates, Principally Fishes.' *American Journal of Science* 282: 1–44.

Olsen, P.E., C.L. Remington, B. Cornet, and K.S. Thomson. 1978. 'Cyclic Change in Late Triassic Lacustrine Communities.' *Science* 201: 729–33.

Orr, W.N. 1986. 'A Norian (Late Triassic) Ichthyosaur from the Martin Bridge Formation, Wallowa Mountains, Oregon.' U.S. Geological Survey Professional Paper 1435: 41–7.

Padian, K. (ed.) 1986. *The Beginning of the Age of Dinosaurs*. Cambridge University Press, Cambridge. 378 pp. (See articles by S. Chatterjee, R.A. Long and K. Padian, P.A. Murray, P.E. Olsen and D. Baird, and K. Padian.)

Parrish, J.M. 1986. 'Locomotor Adaptations in the Hindlimb and Pelvis of the Thecodontia.' *Hunteria* 1 (2): 1–35.

Parrish, J.T., and R.L. Curtis. 1982. 'Atmospheric Circulation, Upwelling, and Organic-Rich Rocks in the Mesozoic and Cenozoic Eras.' *Palaeogeography, Palaeoclimatology, Palaeoecology* 40: 31–66.

Parrish, J.M., and M.G. Lockley. 1984. 'Dinosaur Trackways from the Triassic of Western Colorado.' Geological Society of America, Abstracts with Programs 16 (4): 250.

Parrish, J.M., and R.A. Long. 1983. 'Vertebrate Paleontology of the Late Triassic Chinle Formation, Petrified Forest and vicinity, Arizona.' Geological Society of America, Rocky Mountain and Cordilleran Sections, Abstracts with Programs, p. 285.

Parrish, J.T., A.M. Ziegler, and C.R. Scotese. 1982. 'Rainfall Patterns and the Distribution of Coals and Evaporites in the

Mesozoic and Cenozoic.' *Palaeogeography, Palaeoclimatology, Palaeoecology* 49: 67–101.

Picard, M.D. 1978. 'Stratigraphy of Triassic Rocks in West-Central Wyoming.' *Wyoming Geological Association, 30th Annual Field Conference*, pp. 101–30.

Reif, W.E., and F. Westphal. 1984. *Third Symposium on Mesozoic Terrestrial Ecosystems*. Attempto Verlag, Tubingen. 259 pp. (See article by Weishampel.)

Retallack, G.J. 1977. 'Reconstructing Triassic Vegetation of Eastern Australasia: A New Approach for the Biostratigraphy of Gondwanaland.' *Alcheringa* 1: 247–77.

Schaeffer, B. 1967. 'Late Triassic Fishes from the Western United States.' *Bulletin of the American Museum of Natural History* 135: 285–342.

Sigogneau-Russell, D. 1983. 'Nouveaux taxons de mammifères rhétiens. *Acta Palaeontologica Polonica* 28: 233–49.

Smiley, T.L. 1985. 'The Geology and Climate of the Indigenous Forest, Petrified Forest National Park, Arizona.' *Museum of Northern Arizona Bulletin* 54: 9–15.

Stewart, J.H., F.G. Poole, and R.F. Wilson. 1972. *Stratigraphy and Origin of the Chinle Formation and Related Upper Triassic Strata in the Colorado Plateau Region*. U.S. Geological Survey Professional Paper 690. 336 pp.

Tozer, E.T. 1982. 'Marine Triassic Faunas of North America: Their Significance for Assessing Plate and Terrane Movements.' *Geologische Rundschau* 71: 1077–1104.

Weber, R. 1980. 'Megafosiles de coniferas del Triasico Tardio y del Cretacico Tardio de Mexico y consideraciones generales sobre las coniferas Mesozoicas de Mexico.' Univ. Nac. Auton. Mexico, Inst. Geologia, *Revista* 4: 111–24 (see also pp. 125–73).

Weems, R.E. 1980. 'An Unusual Newly Discovered Archosaur from the Upper Triassic of Virginia, U.S.A.' *Transactions of the American Philosophical Society* 70: 1–53.

Welles, S.P. 1984. '*Dilophosaurus witherilli* (Dinosauria, Theropoda) Osteology and Comparisons.' *Palaeontographica Abt.* A 185: 85–180.

White, G.W. 1980. 'Permian-Triassic Continental Reconstruction of the Gulf of Mexico–Caribbean Area.' *Nature* 283: 823–6.

Wild, R. 1980. 'The Fossil Deposits of Kupferzell, Southwest Germany.' *Mesozoic Vertebrate Life* 1: 15–18.

~ 1983. 'Uber den Ursprung der Flugsaurier.' *Weltenburger Akademie, Erwin Rutte-Festschrift*: 231–8.

The Jurassic

(CHAPTERS THREE AND FOUR)

American Museum of Natural History, field records.

Anderson, R.Y., and D.W. Kirkland. 1960. 'Origin, Varves, and Cycles of Jurassic Todilto Formation, New Mexico.' *American Association of Petroleum Geologists Bulletin* 44: 37–52.

Anonymous. 1981. 'Mid-Jurassic Rocks Cored off Florida.' *Geotimes* 26: 15–16.

Arnold, C.A. 1962. 'A *Rhexoxylon*-like Stem from the Morrison Formation of Utah.' *American Journal of Botany* 49: 883–6.

Attridge, J., A.W. Crompton, and F.A. Jenkins. 1985. 'The Southern African Liassic Prosauropod *Massospondylus* Discovered in North America.' *Journal of Vertebrate Paleontology* 5: 128–32.

*Bakker, R.T. 1980. 'Dinosaur Heresy – Dinosaur Renaissance: Why We Need Endothermic Archosaurs.' *American Association for the Advancement of Science, Selected Symposium* 28: 351–462.

Barthel, K.W. 1978. *Solnhofen, ein Blick in die Erdgeschichte*. Ott Verlag, Thun, Switzerland. 393 pp.

Bell, W.A. 1956. *Lower Cretaceous Floras of Western Canada*. Geological Survey of Canada Memoir 285. 329 pp.

*Benton, M.J. 1984. 'Dinosaurs' Lucky Break.' *Natural History* 1984 (6): 53–9.

Berman, D.S., and J.S. McIntosh. 1978. *Skull and Relationships of the Upper Jurassic Sauropod* Apatosaurus (*Reptilia, Saurischia*). Carnegie Museum of Natural History Bulletin 8. 35 pp.

Bird, R.T. 1985. (See references for Preface.)

Bonaparte, J.F. 1978. *El Mesozoico de America del Sur y sus tetrapodos*. Opera Lilloana 26. 596 pp.

~ 1986. 'Les dinosaures (Carnosaures, Allosaurides, Sauropodes, Cetiosaurides) du Jurassic Moyen de Cerro Condor (Chubut, Argentine).' *Annales de Paleontologie* 72: 247–89, 325–86.

Brown, B. 1941. 'The Age of Sauropod Dinosaurs.' *Science* 93: 594–5.

Brown, J.T. 1972. 'The Upper Jurassic Flora of the Morrison Formation from Central Montana.' *American Journal of Botany* 59: 659.

Buffrenil, V. de, J.O. Farlow, and A. de Ricqles. 1986. 'Growth and Function of *Stegosaurus* Plates: Evidence from Bone Histology.' *Paleobiology* 12: 459–73.

Busbey, A.B., and C. Gow. 1984. 'A New Protosuchian Crocodile from the Upper Triassic Elliot Formation of South Africa.' *Palaeontologia Africana* 25: 127–49.

Callison, G., and H.M. Quimby. 1984. 'Tiny Dinosaurs: Are They Fully Grown?' *Journal of Vertebrate Paleontology* 3: 200–9.

Camp, C.L. 1936. 'A New Type of Small Bipedal Dinosaur from the Navajo Sandstone of Arizona.' *University of California Publications in Geological Sciences* 24: 39–56.

Chandler, M.E.J., and J.A. Jensen. 1966. 'Fruiting Organs from the Morrison Formation of Utah, U.S.A.' *British Museum (Natural History) Bulletin* 12: 139–72.

Coe, M.J., D.L. Dilcher, J.O. Farlow, D.M. Jarzen, and D.A. Russell. 1987. 'Dinosaurs and Land Plants.' In E.M. Friis, W. Chaloner, and P. Crane (eds.), *The Origins of Angiosperms and Their Biological Consequences*, pp. 225–58. Cambridge University Press, Cambridge.

Colbert, E.H. 1968. (See references for Preface.)

~ 1969. *A Jurassic Pterosaur from Cuba*. American Museum Novitates Number 2370. 26 pp.

~ 1981. 'A Primitive Ornithischian Dinosaur from the Kayenta Formation of Arizona.' *Museum of Northern Arizona Bulletin* 53: 1–61.

Colbert, E.H., and C.C. Mook. 1951. 'The Ancestral Crocodilian *Protosuchus*.' *American Museum of Natural History Bulletin* 97: 143–82.

Coombs, W.P., Jr. 1978. 'The Families of the Ornithischian Dinosaur Order Ankylosauria.' *Palaeontology* 21: 143–70.

~ 1980. 'Swimming Ability of Carnivorous Dinosaurs.' *Science* 207: 1198–9.

Cooper, M.R. 1981. 'The Prosauropod Dinosaur *Massospondylus carinatus* Owen from Zimbabwe.' *Occasional Papers of the National Museums and Monuments (Zimbabwe)* 6: 689–840.

~ 1984. 'A Reassessment of *Vulcanodon karibaensis* (Dinosauria: Saurischia) and the Origin of the Sauropoda.' *Palaeontologia Africana* 25: 203–31.

Cousiminer, H.L. 1978. 'Palynofloras of Jurassic Circum-gulf Salt: Age, Environment and Source.' *Palynology* 2: 216.

Craig, L.C., C.N. Holmes, R.A. Cadigen, V.L. Freeman, T.E. Mullens, and G.W. Weir. 1955. 'Stratigraphy of the Morrison and Related Formations, Colorado Plateau region, a Preliminary Report.' *Bulletin U.S. Geological Survey* 1009E: 125–66.

Crush, P.J. 1984. 'A Late Upper Triassic Sphenosuchid Crocodilian from Wales.' *Palaeontology* 27: 131–57.

De Ricqles, A.J. 1983. 'Cyclical Growth in the Long Limb Bones of a Sauropod Dinosaur.' *Acta Palaeontologica Polonica* 28: 225–32.

Delevoryas, T., and R.C. Hope. 1976. 'More Evidence for a Slender Growth Habit in Mesozoic Cycadophytes.' *Review of Palaeobotany and Palynology* 21: 93–100.

Delevoryas, T., and S.C. Srivastava. 1981. 'Jurassic Plants from the Department of Fanscisco Morazan, Central Honduras.' *Review of Palaeobotany and Palynology* 34: 345–57.

Derman, A.S., B.H. Wilkenson, and J.A. Dorr, Jr. 1984. 'Jurassic-Cretaceous Nonmarine Foreland Basin Sedimentation in Western United States.' *American Association of Petroleum Geologists Bulletin* 68: 470–1.

Desmond, A. 1982. *Archetypes and Ancestors: Palaeontology in Victorian London 1850–1875.* University of Chicago Press, Chicago. 287 pp.

Dodson, P. 1980. 'Comparative Osteology of the American Ornithopods *Camptosaurus* and *Tenontosaurus*.' *Mémoires de la Société géologique de France* 139: 81–5.

Dodson, P., A.K. Behrensmeyer, and R.T. Bakker. 1980. 'Taphonomy of the Morrison Formation (Kimmmeridgian-Portlandian) and Cloverly Formation (Aptian-Albian) of the Western United States.' *Mémoires de la Société géologique de France* 139: 87–93.

Dodson, P., A.K. Behrensmeyer, R.T. Bakker, and J.S. McIntosh. 1980. 'Taphonomy and Paleoecology of the Dinosaur Beds of the Jurassic Morrison Formation.' *Paleobiology* 6: 208–32.

Dong, Z., and Z. Tang. 1983. 'Note on the New Mid-Jurassic Ornithopod from Sichuan Basin, China.' *Vertebrata Palasiatica* 21: 168–73.

Dong, Z., and Z. Tang. 1984. 'Note on a New Mid-Jurassic Sauropod (*Datousaurus bashanensis gen. et sp. nov.*) from Sichuan Basin, China.' *Vertebrate Palasiatica* 22: 69–75.

Dong, Z., Z. Tang, and S. Zhou. 1982. 'Report on the Shunosaurian Fauna of Dashanpu, Zigong, Sichuan, Part I, Stegosaurs.' *Vertebrate Palasiatica* 20: 83–8.

Dong, Z., S. Zhou, and Y. Zhang. 1983. *The Dinosaurian Remains from Sichuan Basin, China.* Palaeontologias Sinica, Whole Number 162 Series C, Number 23. 145 pp.

Dutuit, J.M., and A. Ouazzou. 1980. 'Découverte d'une piste de dinosaure sauropode sur le site d'empreintes de Demnat (Haut-Atlas moracain).' *Société géologique de France, Mémoire* 139: 95–102.

Ferrusquia-Villafranca, I., S.P. Applegate, and L. Espinosa-Arrubarrena. 1978. 'Rocas volcanosedimentarias Mesozoicas y huellas de dinosaurios en la region suroccidental Pacifica de Mexico.' *Universidad Nacional Autonoma de Mexico, Instituto de Geologia, Revisto* 2: 150–62.

Gaffney, E.S. 1979. 'The Jurassic Turtles of North America.' *Bulletin of the American Museum of Natural History* 162: 91–136.

Galton, P.M. 1976. *Prosauropod Dinosaurs (Reptilia: Saurischia) of North America.* Postilla Peabody Museum, Yale University, 169. 98 pp.

~ 1978. 'Fabrosauridae, the Basal Family of Ornithischian Dinosaurs.' *Palaontologische Zeitschrift* 52: 138–59.

~ 1980. 'Armored Dinosaurs (Ornithischia: Ankylosauria) from the Middle and Upper Jurassic of England.' *Geobios* 13: 825–37.

~ 1981. '*Dryosaurus*, a Hypsilophodontid Dinosaur from the Upper Jurassic of North America and Africa, Postcranial Skeleton.' *Palaontologische Zeitschrift* 55: 271–312.

~ 1982. '*Elaphrosaurus*, an Ornithomimid Dinosaur from the Upper Jurassic of North America and Africa.' *Palaontologische Zeitschrift* 56: 265–75.

~ 1982. 'Juveniles of the Stegosaurian Dinosaur *Stegosaurus* from the Upper Jurassic of North America.' *Journal of Vertebrate Paleontology* 2: 47–62.

~ 1982. 'The Postcranial Anatomy of Stegosaurian Dinosaur *Kentrosaurus* from the Upper Jurassic of Tanzania, East Africa.' *Geologica et Palaeontologica* 15: 139–60.

~ 1983. 'The Cranial Anatomy of *Dryosaurus*, a Hypsilophodontid Dinosaur from the Upper Jurassic of North America and East Africa, with a Review of Hypsilophodontids from the Upper Jurassic of North America.' *Geologica et Palaeontologica* 17: 207–43.

Galton, P.M., R. Brun, and M. Rioult. 1980. 'Skeleton of the Stegosaurian Dinosaur *Lexovisaurus* from the Lower Part of the Middle Callovian (Middle Jurassic) of Argences (Calvados), Normandy.' *Bull. trim. Soc. géol. Normandie et Amis Museum du Havre* 67: 39–53.

Galton, P.M., and J.A. Jensen. 1973. 'Skeleton of a Hypsilophodontid Dinosaur (*Nanosaurus(?) rex*) from the Upper Jurassic of Utah.' *Brigham Young University, Geological Studies* 20: 137–57.

Galton, P.M., and J.A. Jensen. 1979. 'A New Large Theropod Dinosaur from the Upper Jurassic of Colorado.' *Brigham Young University, Geological Studies* 26: 1–12.

Galton, P.M., and H.P. Powell. 1980. 'The Ornithischian Dinosaur *Camptosaurus prestwitchii* from the Upper Jurassic of England.' *Palaeontology* 23: 411–43.

Galton, P.M., and H.P. Powell. 1983. 'Stegosaurian Dinosaurs from the Bathonian (Middle Jurassic) of England, the earliest Record of the Family Stegosauridae.' *Geobios* 16: 219–29.

Gilmore, C.W. 1909. 'Osteology of the Jurassic Reptile *Camptosaurus*, with a Revision of the Species of the Genus, and Description of Two New species.' *Proceedings of the U.S. National Museum* 36: 195–332.

~ 1914. *Osteology of the Armored Dinosauria in the United States National Museum, with Special Reference to the Genus Stegosaurus.* U.S. National Museum Bulletin 89. 136 pp.

~ 1920. *Osteology of the Carnivorous Dinosauria in the United States National Museum, with Special Reference to the Genera* Antrodemus (Allosaurus) *and* Ceratosaurus. U.S. National Museum Bulletin 110. 154 p.

~ 1925. 'A Nearly Complete Articulated Skeleton of *Camarasaurus*, a Saurischian Dinosaur from the Dinosaur National Monument, Utah.' *Carnegie Museum of Natural History Memoirs* 10: 347–84.

~ 1932. 'On a Newly Mounted Skeleton of *Diplodocus* in the United States National Museum.' *Proceedings of the U.S. National Museum* 81: 1–21.

~ 1936. 'Osteology of *Apatosaurus*, with Special Reference to Specimens in the Carnegie Museum.' *Carnegie Museum of Natural History Memoirs* 11: 175–300.

Green, M.W. 1975. 'Paleodepositional Units in Upper Jurassic Rocks in Gallup-Laguna Area, New Mexico.' *American Association of Petroleum Geologists, Bulletin* 59: 910.

Greive, R.A.F. 1982. 'The Record of Impact on Earth: Implications for a Major Cretaceous/Tertiary Impact Event.' *Geological Society of America Special Paper* 190: 25–37.

*Hallam, A. 1981. 'The End-Triassic Bivalve Extinction Event.' *Palaeogeography, Palaeoclimatology, Palaeoecology* 35: 1–44.

~ 1983. 'Early and Mid-Jurassic Molluscan Biogeography and the Establishment of the Central Atlantic Seaway.' *Palaeogeography, Palaeoclimatology, Palaeoecology* 43: 181–93.

Hamblin, A.P., and R.G. Walker. 1979. 'Storm Dominated Shallow Marine Deposits: The Fernie-Kootenay (Jurassic) Transition.' *Canadian Journal of Earth Sciences* 16: 1673–90.

Harrison, J.C., Q.H. Goodbody, and R.L. Christie. 1985. 'Stratigraphic and Structural Studies on Melville Island, District of Franklin.' Geological Survey of Canada Paper 85-1A: 629–37.

Harshbarger, J.W., C.A. Repenning, and J.H. Irwin. 1957. *Stratigraphy of the Uppermost Triassic and the Jurassic Rocks of the Navajo Country.* U.S. Geological Survey Professional Paper 291. 74 pp.

Hatcher, J.B. 1901. '*Diplodocus* (Marsh): Its Osteology, Taxonomy, and Probable Habits, with a Restoration of the Skeleton.' *Carnegie Museum of Natural History Memoirs* 1: 1–63.

~ 1903. 'Osteology of *Haplocanthosaurus*, with Description of New Species and Remarks on Probable Habits of the Sauropoda, and the Age and Origin of the *Atlantosaurus* Beds.' *Carnegie Museum of*

Natural History Memoirs 2: 1–75.

Hay, W.W., J.F. Behensky, E.J. Barron, and J.L. Sloan. 1982. 'Late Triassic-Liassic Paleoclimatology of the Proto-Central North Atlantic Rift System.' *Palaeogeography, Palaeoclimatology, Palaeoecology* 40: 13–30.

He, X., and K. Cai. 1983. 'A New Species of *Yandusaurus* (Hypsilophodont Dinosaur) from the Middle Jurassic of Dashanpu, Zigong, Sichuan.' *Journal of the Chengdu College of Geology, Supplement* 1: 5–14.

Hecht, M.K., J.H. Ostrom, G. Viohl, and P. Wellnhofer (eds.). 1985. *The Beginnings of Birds*. Eichstatt, West Germany, Proceedings of the International *Archaeopteryx* Conference. 367+ pp. (See articles by J. Gauthier and K. Padian and by J.H. Ostrom.)

Holland, W.J. 1906. 'The Osteology of *Diplodocus.' Carnegie Museum of Natural History Memoirs* 2: 225–64.

~ 1924. 'The Skull of *Diplodocus.' Carnegie Museum of Natural History Memoirs* 9: 397–403.

Hubert, J.F., J.M. Gilchrist, and A.A. Reed. 1982. 'Jurassic Redbeds of the Connecticut Valley.' *State Geological and Natural History Survey of Connecticut, Guidebook* 5: 103–41.

Hubert, J.F., and K.A. Mertz. 1980. 'Eolian Dune Field of Late Triassic Age, Fundy Basin, Nova Scotia.' *Geology* 8: 516–19. (See also comment, P.E. Olsen, *Geology* 9: 557–9.)

Huene, F. von. 1926. 'Vollstandige Osteologie eines Plateosauriden aus dem Schwabischen Keuper.' *Geologische und Palaeontologische Abhandlungen, Neue Folge* 15: 139–79.

Imlay, R.W. 1980. *Jurassic Paleobiogeography of the Conterminous United States in Its Continental Setting*. U.S. Geological Survey Professional Paper 1062. 134 pp.

Jacobs, L.L. (ed.). 1980. (See references for Triassic.) (See articles by D. Baird, J.F. Bonaparte, A.W. Crompton and K.K. Smith, S.L. Jain, P.E. Olsen, and J.H. Ostrom.)

Jansa, L.F., A.P. Hamelin, and R.G. Walker. 1979. 'Storm-Dominated Shallow Marine Deposits: The Fernie-Kootenay (Jurassic) Transition, Southern Rocky Mountains.' *Canadian Journal of Earth Sciences* 16: 1673–90.

Jensen, J.A. 1981. 'Another Look at *Archaeopteryx* as the World's Oldest Bird.' *Encyclia* 58: 109–28.

~ 1984. 'Continuing Studies of New Jurassic/Cretaceous Vertebrate Faunas from Colorado and Utah.' *National Geographic Research Reports* 16: 373–81.

~ 1985. 'Three New Sauropod Dinosaurs from the Upper Jurassic of Colorado.' *Great Basin Naturalist* 45: 697–709.

Kearney, C. 1958. 'Big Bone.' *Natural History* 67: 113.

Kitching, J.W., and M.A. Raath. 1984. 'Fossils from the Elliot and Clarens Formations (Karoo Sequence) of the Northeastern Cape, Orange Free State and Lesotho, and a Suggested Biozonation Based on Tetrapods.' *Palaeontologia Africana* 25: 111–25.

Kocurek, G. 1981. 'Erg Reconstruction: The Entrada Sandstone (Jurassic) of Northern Utah and Colorado.' *Palaeogeography, Palaeoclimatology, Palaeoecology* 36: 125–53.

Lapparent, A.F. 1955. *Etude paléontologique des vértebrés du Jurassic d'el Mers*. Notes et mémoires, Service Géologique du Maroc 124. 36 pp.

Lockley, M.G., K.J. Houck, and N.K. Prince. 1986. 'North America's Largest Dinosaur Trackway Site: Implications for Morrison Formation Paleoecology.' *Geological Society of America Bulletin* 97: 1163–76.

Lucas, S.G. 1985. 'The Jurassic System in East-Central New Mexico.' *New Mexico Geological Society Guidebook, 36th Field Conference, Santa Rosa*: 213–42.

Lull, R.S. 1919. 'The Sauropod Dinosaur *Barosaurus* Marsh.' *Memoirs of the Connecticut Academy of Arts and Letters* 6: 1–42.

McDonald, N.G. 1982. 'Paleontology of the Mesozoic Rocks of the Connecticut Valley.' *State Geological and Natural History Survey of Connecticut, Guidebook* 5: 143–72.

McGowan, C. 1978. 'Further Evidence for the Wide Geographical Distribution of Ichthyosaur Taxa (Reptilia: Ichthyosauria).' *Journal of Paleontology* 52: 1144–62.

McGowan, C. 1983. (See references for Preface.)

McIntosh, J.S. 1977. *Dinosaur National Monument*. Constellation, Phoenix. 41 pp.

~ 1981. *Annotated Catalogue of the Dinosaurs (Reptilia, Archosauria) in the Collections of Carnegie Museum of Natural History*. Carnegie Museum of Natural History Bulletin 18. 67 pp.

McIntosh, J.S., and D.S. Berman. 1975. 'Description of the Palate and Lower Jaw of the Sauropod Dinosaur *Diplodocus* (Reptilia: Saurischia) with Remarks on the Nature of the Skull of *Apatosaurus.' Journal of Paleontology* 49: 187–99.

McKee, E.D. 1979. 'Ancient Sandstones Considered to be Aeolian.' U.S. Geological Survey Professional Paper 1052: 187–233.

Madsen, J.H., Jr. 1974. 'A New Theropod Dinosaur from the Upper Jurassic of Utah.' *Journal of Paleontology* 48: 27–31.

~ 1976. Allosaurus fragilis: *Revised Osteology*. Utah Department of Natural Resources Bulletin 109. 163 pp.

~ 1976. 'A Second New Theropod Dinosaur

from the Late Jurassic of East Central Utah.' *Utah Geology* 3: 51–60.

Mallory, W.W. (ed.). 1972. *Geologic Atlas of the Rocky Mountain Region*. Rocky Mountain Association of Geologists, Denver. 331 pp. (See article by J.A. Peterson.)

Marsh, O.C. 1896. 'Dinosaurs of North America.' *Annual Report, U.S. Geological Survey*, part I, 16: 133–244.

Marzolf, J.E. 1984. 'Changing Wind and Hydrologic Regimes during Deposition of the Navajo and Aztec Sandstones, Jurassic (?), Southwestern United States.' *Developments in Sedimentology* 38: 635–60.

Matthew, W.D. 1915. *Dinosaurs, with Special Reference to the American Museum Collections*. American Museum of Natural History Handbook 5. 117 pp.

Middleton, L.T., and R.C. Blakey. 1983. 'Facies Associations and Controls on Jurassic Navajo-Kayenta Intertonguing, Northern Arizona.' Geological Society of America, Rocky Mountain and Cordilleran Sections, Abstracts with Programs, p. 286.

Moberly, R. 1960. 'Morrison, Cloverly, and Sykes Mountain Formations, Northern Bighorn Basin, Wyoming and Montana.' *Bulletin Geological Society of America* 71: 1137–76.

Molnar, R.E. 1982. 'Australian Mesozoic Reptiles.' In P.V. Rich and E.M. Thompson (eds.), *The Fossil Vertebrate Record of Australasia*, pp. 169–225. Monash University, Melbourne.

Mook, C.C. 1916. 'A Study of the Morrison Formation.' *Annals of the New York Academy of Sciences* 17: 39–191.

~ 1917. 'The Fore and Hind Limbs of *Diplodocus.' American Museum of Natural History Bulletin* 37: 815–19.

Morales, M. 1986. 'Dinosaur Tracks in the Lower Jurassic Kayenta Formation near Tuba City, Arizona.' In M. Lockley, *A Guide to Dinosaur Tracksites of the Colorado Plateau and the American Southwest*, University of Colorado at Denver Geology Department Magazine Special Issue 1: 14–19.

Nicholls, E.L. 1976. 'The Oldest Known North American Occurrence of the Plesiosauria (Reptilia: Sauropterygia) from the Liassic (Lower Jurassic) Fernie Group, Alberta, Canada.' *Canadian Journal of Earth Sciences* 13: 185–8.

Olsen, P.E. 1986. 'Discovery of Earliest Jurassic Reptile Assemblages from Nova Scotia Imply Catastrophic End to the Triassic.' *Lamont Newsletter* 12: 1–3.

~ 1986. 'Impact Theory: Is the Past the Key to the Future?' *Lamont-Doherty Geological Observatory of Columbia University, 1983–1986, pp. 5–10*. Lamont-Doherty Geological Observatory, Palisades, New York.

Olsen, P.E., and D. Baird. 1982. 'Early Jurassic Vertebrate Assemblages from the McKoy Brook Formation of the Fundy Group (Newark Supergroup, Nova Scotia).' Geological Society of America, Abstracts with Program 14 (1–2): 70.

Olsen, P.E., and P.M. Galton. 1984. 'A Review of the Reptile and Amphibian Assemblages from the Stormberg of Southern Africa, with Special Emphasis on the Footprints and the Age of the Stormberg.' *Palaeontologia Africana* 25: 87–110.

Olsen, P.E., A.M. McCune, and K.S. Thompson. 1982. 'Correlation of the Early Mesozoic Newark Supergroup by Vertebrates, Principally Fishes.' *American Journal of Science* 282: 1–44.

Osborn, H.F. 1899. 'A Skeleton of *Diplodocus*.' *Memoirs of the American Museum of Natural History* 1: 191–214.

~ 1904. 'Manus, Sacrum and Caudals of Sauropoda.' *Bulletin of the American Museum of Natural History* 20: 181–90.

~ 1906. 'The Skeleton of *Brontosaurus* and the Skull of *Morosaurus*.' *Nature* 73: 282–4.

~ 1917. 'Skeletal Adaptations of *Ornitholestes, Struthiomimus*, and *Tyrannosaurus*.' *Bulletin of the American Museum of Natural History* 35: 733–71.

Osborn, H.F., and C.C. Mook. 1921. '*Camarasaurus, Amphicoelias*, and Other Sauropods of Cope.' *American Museum of Natural History Memoirs* 3: 247–387.

Ostrom, J.H. 1972. 'Were Some Dinosaurs Gregarious?' *Palaeogeography, Palaeoclimatology, Palaeoecology* 11: 287–301.

~ 1978. 'A New Look at Dinosaurs.' *National Geographic* 154 (2): 152–85.

Ostrom, J.H., and J.S. McIntosh. 1966. *Marsh's Dinosaurs: The Collections from Como Bluff*. Yale University Press, New Haven. 388 pp.

Owen, R. 1849-84. *A History of British Fossil Reptiles*. 4 volumes. London, Cassell and Company.

Padian, K. 1984. 'Pterosaur Remains from the Kayenta Formation (? Early Jurassic) of Arizona.' *Palaeontology* 27: 407–13.

~ 1986. (See references for Triassic.) (See articles by *M.J. Benton, J.F. Bonaparte, J.M. Clark and D.E. Fastovksy, A.W. Crompton and J. Attridge, and P.E. Olsen and H.D. Suess.)

Padian, K., and P.E. Olsen. 1984. 'The Fossil Trackway *Pteraichnus*: Not Pterosaurian but Crocodilian.' *Journal of Paleontology* 58: 178–84.

Person, C.P., and T. Delevoryas. 1982. 'The Middle Jurassic Flora of Oaxaca Mexico.' *Palaeontographica Abt.* B 180: 82–119.

Peterson, F. 1984. 'Fluvial Sedimentation on a Quivering Craton.' *Sedimentary Geology* 38: 21–49.

Peterson, F., and G.N. Pipiringos. 1979. *Stratigraphic Relations of the Navajo Sandstone to Middle Jurassic Formations, Southern Utah and Northern Arizona*. U.S. Geological Survey Professional Paper 1035B. 43 pp.

Peterson, O.A., and C.W. Gilmore. 1902. '*Elosaurus parvus*: A New Genus and Species of the Sauropoda.' *Annals of the Carnegie Museum* 1: 490–9.

Powell, T.G. 1985. 'Paleogeographic Implications for the Distribution of Upper Jurassic Source Beds: Offshore Eastern Canada.' *Bulletin of Canadian Petroleum Geology* 33: 116–19.

Prothero, D.R. 1981. 'New Jurassic Mammals from Como Bluff, Wyoming, and the Interrelationships of Non-tribosphenic Theria.' *Bulletin of the American Museum of Natural History 167*: 277–326.

Prothero, D.R., and R. Estes. 1980. 'Late Jurassic Lizards from Como Bluff, Wyoming and Their Palaeobiogeographic Significance.' *Nature* 286: 484–6.

Rasmussen, T.E., and G. Callison. 1981. 'A New Species of Triconodont Mammal from the Upper Jurassic of Colorado.' *Journal of Paleontology* 55: 628–34.

Raup, D.M., and J.J. Sepkoski. 1982. (See references for 'Before the Mesozoic.')

Rigby, J.K. 1982. '*Camarasaurus* Cf. *supremus* from the Morrison Formation near San Ysidro, New Mexico – the San Ysidro Dinosaur.' *New Mexico Geological Society Guidebook, 33rd Field Conference, Albuquerque Country* II: 271–2.

Riggs, E.S. 1903. 'Structure and Relationships of Opisthocoelian Dinosaurs, Part I, *Apatosaurus*.' *Field Columbian Museum, Geological Series* 2: 165–96.

Santa Luca, A.P. 1976. 'A Complete Skeleton of the Late Triassic Ornithischian *Heterodontosaurus tucki*.' *Nature* 264: 324–8.

Santa Luca, A.P. 1980. 'The Postcranial Skeleton of *Heterodontosaurus tucki* (Reptilia, Ornithischia) from the Stormberg of South Africa.' *Palaeontologia Africana* 25: 151–80.

Seidemann, D.E., W.D. Masterson, M.P. Dowling, and K.K. Turkekian. 1984. 'K-Ar Dates and 40Ar/39Ar Age Spectra for Mesozoic Basalt Flows of the Hartford Basin, Connecticut, and the Newark Basin, New Jersey.' *Geological Society of America Bulletin* 95: 594–8.

Silva Pineda, A. 1982. 'Jurassic Cycadophytes of Mexico.' *Third North American Paleontologiucal Convention Proceedings* 2: 489–94.

Simpson, G.G. 1942. (See references for Preface.)

Soreno, P. 1986. 'Phylogeny of the Bird-hipped Dinosaurs (Order Ornithischia).' *National Geographic Research* 2: 234–56.

Stokes, W.L. 1978. 'Animal Tracks in the Navajo-Nugget Sandstone.' *University of Wyoming Contributions to Geology* 16: 103–7.

~ 1983. 'Silicified Trees in the Navajo Sandstone, East-Central Utah.' Geological Society of America, Rocky Mountain and Cordilleran Sections, Abstracts with Program, p. 286.

Stokes, W.L., and J.H. Madsen. 1979. 'Environmental Significance of Pterosaur Tracks in the Navajo Sandstone (Jurassic), Grand County, Utah.' *Brigham Young University Geology Studies* 26: 21–6.

Stovall, J.W. 1938. 'The Morrison of Oklahoma and Its Dinosaurs.' *Journal of Geology* 46: 583–600.

Surlyk, F., J.H. Callomon, R.G. Bromley, and T. Birkelund. 1973. *Stratigraphy of the Jurassic–Lower Cretaceous Sediments of Jameson Land and Scoresby Land, East Greenland*. Meddelelser om Gronland 193 (5). 76 pp.

Szigeti, G.J., and J.E. Fox. 1981. 'Unkpapa Sandstone (Jurassic), Black Hills, South Dakota: An Aeolian Facies of the Morrison Formation.' *Special Publication, Society of Economic Paleontologists and Mineralogists* 31: 331–49.

Thulborn, R.A. 1970. 'The Skull of *Fabrosaurus australis*, a Triassic Ornithischian Dinosaur.' *Palaeontology* 13: 414–32.

~ 1972. 'The Post-Cranial Skeleton of the Triassic Ornithischian Dinosaur *Fabrosaurus australis*.' *Palaeontology* 15: 29–60.

Thurston, H. 1986. 'The Fossils of Fundy.' *Atlantic Insight*, July: 20–3.

Tozer, E.T. 1971. 'One, Two or Three Connecting Links between Triassic and Jurassic Ammonoids.' *Nature* 232: 565–6.

~ 1971. 'Triassic Time and Ammonoids: Problems and Proposals.' *Canadian Journal of Earth Sciences* 8: 989–1031.

Traverse, A. 1987. 'Pollen and Spores Date Origin of Rift Basins from Texas to Nova Scotia as Early Late Triassic.' *Science* 236: 1469–72.

Tschudy, R.H., B.E. Tschudy, S. Van Loenen, and G. Doher. 1981. *Illustrations of Plant Microfossils from the Morrison Formation.* I *Plant Microfossils from the Brushy Basin Member*. U.S. Geological Survey Open File Report 81–35. 21 pp.

Tucker, M.E., and M.J. Benton. 1982. 'Triassic Environments, Climates and Reptile Evolution.' *Palaeogeography, Palaeoclimatology, Palaeoecology* 40: 361–79.

Van Straaten, L.M.J.U. 1971. 'Origin of Solnhofen Limestone.' *Geologie en*

Mijnbouw 50: 3–8.

Visser, J.N.J. 1984. 'A Review of the Stormberg Group and Drakensberg Volcanics in Southern Africa.' *Palaeontologia Africana* 25: 5–27.

Wadsworth, N. 1973. 'Colorado's 100-Foot Dinosaur: Is It the World's Largest?' *Science Digest*, April: 77–81.

Weaver, J.C. 1983. 'The Improbable Endotherm: The Energetics of the Sauropod Dinosaur *Brachiosaurus*.' *Paleobiology* 9: 173–82.

Weiss, M.P., and H.E. Wenden. 1965. 'Calcite Spherulites from the Morrison Formation, Wyoming.' *Journal of Sedimentary Petrology* 35: 985–8.

Welles, S.P. 1954. 'New Jurassic Dinosaur from the Kayenta Formation of Arizona.' *Geological Society of America Bulletin* 65: 591–8.

~ 1971. 'Dinosaur Footprints from the Kayenta Formation of Northern Arizona.' *Plateau* 44: 27–38.

~ 1984. (See references for Triassic.)

West, L., and D. Chure. 1984. *Dinosaur: The Dinosaur National Monument Quarry.* Dinosaur Nature Association, Jensen, Utah. 40 pp.

West, S. 1979. 'Dinosaur Head Hunt.' *Science News* 116 (18): 314–15.

White, G.W. 1980. (See references for Triassic.)

White, T.E. 1958. 'The Braincase of *Camarasaurus lentus*.' *Journal of Paleontology* 32: 477–94.

Wild, R. 1978. *Ein Sauropoden-Rest aus dem Posidonienschiefer (Lias, Toarcium) von Holtzmaden.* Stuttgarter Beitrage zur Naturkunde, Serie B Nr. 41. 15 pp.

Zhou, S. 1983. 'A Nearly Complete Skeleton of Stegosaur from Middle Jurassic of Dashanpu, Zigong, Sichuan.' *Journal of the Chengdu College of Geology, Supplement* 1: 15–26.

The Early Cretaceous

(CHAPTER FIVE)

Allen, P. 1975. 'Wealden of the Weald: A New Model.' *Proceedings of the Geologists' Association* 86: 389–437.

Alvin, K.L. 1983. 'Reconstruction of a Lower Cretaceous Conifer.' *Botanical Journal of the Linnean Society* 86: 169–76.

Alvin, K.L., C.J. Fraser, and R.A. Spicer. 1981. 'Anatomy and Paleoecology of *Pseudofrenelopsis* and Associated Conifers in the English Wealden.' *Palaeontology* 24: 759–78.

Axelrod, D.I. 1985. 'Reply to Comments on Cretaceous Climatic Equability in Polar Regions.' *Palaeogeography, Palaeoclimatology,*

Palaeoecology 49: 357–9.

Balkwill, H.R. 1983. *Geology of Amund Ringnes, Cornwall, and Haig-Thomas Islands, District of Franklin.* Geological Survey of Canada Memoir 390. 76 pp.

Barbour, E.H. 1931. 'Evidence of Dinosaurs in Nebraska.' *Nebraska State Museum Bulletin* 21: 187–90.

Basinger, J.F., and D.L. Dilcher. 1984. 'Ancient Bisexual Flowers.' *Science* 224: 511–13.

Batten, D.J. 1974. 'Wealden Palaeoecology from the Distribution of Plant Fossils.' *Proceedings of the Geologists' Association* 85: 433–58.

~ 1982. 'Palynofacies and Salinity in the Purbeck and Wealden of Southern England.' In F.T. Banner and A.R. Lord (eds.), *Aspects of Micropalaeontology*, pp. 278–308. Allen and Unwin, London.

Bell, W.A. 1956. (See references for Jurassic.)

Bird, R.T. 1944. 'Did *Brontosaurus* Ever Walk on Land?' *Natural History* 53: 60–7.

~ 1954. 'We Captured a "Live" Brontosaur.' *National Geographic* 105: 707–22.

~ 1985. (See references for Preface.)

Bjork, P.R. 1985. 'A New Iguanodontid Dinosaur from the Lakota Formation of the Northern Black Hills of South Dakota.' *Proceedings of the Pacific Division, American Association for the Advancement of Science* 4: 22.

Blows, W.T. 1982. 'A Preliminary Account of a New Specimen of *Polacanthus foxi* (Ankylosauria, Reptilia) from the Wealden of the Isle of Wight.' *Proceedings of the Isle of Wight Natural History and Archaeological Society* 7: 303–6.

Bodily, N.M. 1969. 'An Armored Dinosaur from the Lower Cretaceous of Utah.' *Brigham Young University Geological Studies* 16: 35–60.

Brown, B. 1941. (See references for Jurassic.)

Casier, E. 1960. *Les iguanodons de Bernissart.* Institut royal des sciences naturelles de Belgique, Brussels. 134 pp.

Charig, A.J., and A.C. Milner. 1986. '*Baryonyx*, a Remarkable New Theropod Dinosaur.' *Nature* 324: 359–61.

Clark, W.B., A.B. Bibbins, and E.W. Berry. 1911. 'The Lower Cretaceous Deposits of Maryland.' *Maryland Geological Survey, Lower Cretaceous:* 23–98.

Conway Morris, S. 1985. 'Polar Forests of the Past.' *Nature* 313: 739.

Coombs, W.P., Jr. 1971. *The Ankylosauria.* Columbia University, PhD, University Microfilms, Ann Arbor. 287 pp.

~ 1978. (See references for Jurassic.)

Crane, P.R., and D.L. Dilcher. 1985. '*Lesqueria*: an early Angiosperm Fruiting Axis from the mid-Cretaceous.' *Annals of*

the *Missouri Botanical Garden* 71: 384–402.

Currie, P.J. 1981. 'Bird Footprints from the Gething Formation (Aptian, Lower Cretaceous) of Northeastern British Columbia, Canada.' *Journal of Vertebrate Paleontology* 1: 257–64.

~ 1983. 'Hadrosaur Trackways from the Lower Cretaceous of Canada.' *Acta Palaeontologica Polonica* 28: 63–73.

~ 1984. 'I dinosauri del Canada.' In J.F. Bonaparte, E.H. Colbert, P.J. Currie, A. de Rigles, Z. Kielan-Jawarowska, G. Leonardi, N. Morello, and P. Taquet, *Sulle Orme dei Dinosauri*, pp. 111–24. Erizzo Editrice, Venice.

Currie, P.J., and W.A.S. Sarjeant. 1979. 'Lower Cretaceous Dinosaur Footprints from the Peace River Canyon, British Columbia.' *Palaeogeography, Palaeoclimatology, Palaeoecology* 28: 103–15.

Dilcher, D.L. 1984. 'In Pursuit of the First Flower.' *Natural History* 93: 56–61.

~ 1984. 'In Search of Fossil Plants.' *Terra* January-February: 10–16.

Dilcher, D.L., and Peter R. Crane. 1984. '*Archaeanthus*: An Early Angiosperm from the Cenomanian of the Western Interior of North America.' *Annals of the Missouri Botanical Garden* 71: 351–83.

Dodson, P., A.K. Behrensmeyer, and R.T. Bakker. 1980 (See references for Jurassic.)

Dorr, J.A., Jr. 1985. 'Newfound Early Cretaceous Dinosaurs and Other Fossils in Southeastern Idaho and Westermost Wyoming.' *Contributions from the Museum of Paleontology, University of Michigan* 27: 73–85.

Durkee, S.K., and S. Hollis. 1980. 'Depositional Environments of the Lower Cretaceous Smiths Formation within a Portion of the Idaho-Wyoming Thrust Belt.' *Wyoming Geological Associaiton Guidebook* 31: 101–16.

Eaton, T.H., Jr. 1960. A New Armored Dinosaur from the Cretaceous of Kansas.' *University of Kansas Paleontological Contributions Vertebrata* 8: 1–24.

Edwards, M.B., R. Edwards, and E.H. Colbert. 1978. 'Carnosaurian Footprints in the Lower Cretaceous of Eastern Spitsbergen.' *Journal of Paleontology* 52: 940–1.

Elzanowski, A. 1983. 'Birds in Cretaceous Ecosystems.' *Acta Palaeontologica Polonica* 28: 75–92.

Farlow, J.O. 1981. 'Estimates of Dinosaur Speeds from a New Trackway Site in Texas.' *Nature* 294: 747–8.

Fifield, R. 1984. 'When Subtropical England Was Abuzz with Insects.' *New Scientist*, 11 October: 19.

Fontaine, W.M. 1893. 'Notes on Some fossil Plants from the Trinity Division of the

Comanche Series of Texas.' *Proceedings United States National Museum* 16: 261–82.

Forster, C.A. 1984. 'The Paleoecology of the Ornithopod Dinosaur *Tenontosaurus tilletti* from the Cloverly Formation, Big Horn Basin of Wyoming and Montana.' *Mosasaur* 2: 151–63.

Fouch, T.D., J.H. Hanley, and R.M. Forester. 1979. 'Preliminary Correlation of Cretaceous and Paleogene Lacustrine and Related Nonmarine Sedimentary and Volcanic Rocks in Parts of the Eastern Great Basin of Nevada and Utah.' In G.M. Newman and H.D. Goode (eds.) *Basin and Range Symposium, Rocky Mountain Association of Geologists and Utah Geological Association*, pp. 305–12.

Galton, P.M. 1971. 'A Primitive Dome-Headed Dinosaur (Ornithischia-Pachycephalosauridae) from the Lower Cretaceous of England and the Function of the Dome of Pachycephalosaurids.' *Journal of Paleontology* 45: 40–7.

~ 1974. 'The Ornithischian Dinosaur *Hypsilophodon* from the Wealden of the Isle of Wight.' *Bulletin of the British Museum (Natural History) Geology* 25: 1–152.

Galton, P.M., and J.A. Jensen. 1979. 'Remains of Ornithopod Dinosaurs from the Lower Cretaceous of North America.' *Brigham Young University, Geological Studies* 25: 1–10.

Galton, P.M., and P. Taquet. 1982. '*Valdosaurus*: A Hypsilophodontid Dinosaur from the Lower Cretaceous of Europe and Africa.' *Geobios* 15: 147–59.

Gillette, D.D., and D.A. Thomas. 1985. 'Dinosaur Tracks in the Dakota Formation (Aptian-Albian) at Clayton Lake State Park, Union County, New Mexico.' *New Mexico Geological Society Guidebook, 36th Field Conference*: 283–8.

Gilmore, C.W. 1909. (See references for Jurassic.)

~ 1914. (See references for Jurassic.)

~ 1921. 'The Fauna of the Arundel Formation of Maryland.' *Proceedings of the U.S. National Museum* 59: 581–94.

Gries, J.P. 1962. 'Lower Cretaceous Stratigraphy of South Dakota and the Eastern Edge of the Powder River Basin.' In *Symposium on Early Cretaceous Rocks of Wyoming, Wyoming Geological Association Guidebook*: 163–72.

Harris, T.M. 1981. 'Burnt Ferns from the English Wealden.' *Proceedings of the Geologists' Association* 92: 47–58.

Hayes, P.T. 1970. *Cretaceous Paleogeography of southeastern Arizona and Adjacent Areas.* U.S. Geological Survey Professional Paper 658B. 42 pp.

Hickey, L.J., and J.A. Doyle. 1977. 'Early Cretaceous Fossil Evidence for Angiosperm Evolution.' *Botanical Review*

43: 3–104.

Hopkins, W.S., Jr. 1971. 'Palynology of the Lower Cretaceous Isachsen Formation on Melville Island, District of Franklin.' *Geological Survey of Canada Bulletin* 197: 109–33.

Hughes, N.F. 1975. 'Plant Succession in the English Wealden Strata.' *Proceedings of the Geologists' Association* 86: 439–55.

Jacobs, L.L. (ed.). 1980. (See references for Triassic.) (See articles by A.J. Charig and D.A. Russell.)

Jefferson, T.H. 1982. 'Fossil Forests from the Lower Cretaceous of Alexander Island, Antarctica.' *Palaeontology* 25: 671–708.

Jeletzky, J.A. 1971. *Marine Cretaceous Biotic Provinces and Paleogeography of Western and Arctic Canada: Illustrated by a Detailed Study of Ammonites.* Geological Survey of Canada Paper 70–22. 92 pp.

Jensen, J.A. 1984. (See references for Jurassic.)

Kauffman, E.G. 1984. *Paleobiogeography and Evolutionary Response Dynamic in the Cretaceous Western Interior Seaway of North America.* Geological Association of Canada Special Paper 27: 273–306.

Kool, R. 1981. 'The Walking Speed of Dinosaurs from the Peace River Canyon, British Columbia, Canada.' *Canadian Journal of Earth Sciences* 18: 823–5.

Langston, W., Jr. 1974. 'Nonmammalian Comanchean Tetrapods.' *Geoscience and Man* 8: 77–102.

~ 1979. 'Lower Cretaceous Dinosaur Tracks near Glen Rose, Texas.' *Field Trip Guide, American Association of Stratigraphic Palynologists 12th Annual Meeting, Dallas*: 39–55.

Lapasha, C.A., and C.N. Miller. 1984. 'Flora of the Early Cretaceous Kootenai Formation in Montana, Paleoecology.' *Palaeontographica Abt.* B 194: 109–30.

Legault, J.A., and G. Norris. 1982. 'Palynological Evidence for Recycling of Upper Devonian into Lower Cretaceous of the Moose River Basin, James Bay Lowland, Ontario.' *Canadian Journal of Earth Sciences* 19: 1–7.

Lesquereux, L. 1892. *The Flora of the Dakota Group.* U.S. Geological Survey Monograph 17. 400 pp.

Lock, B.E., B.K. Darling, and I.D. Rex. 1983. 'Marginal Marine Evaporites, Lower Cretaceous of Arkansas.' *Bulletin American Association of Petroleum Geologists* 67: 1467–8.

Lockley, M.G. 1985. 'Dinosaur Footprints from the Dakota Group of Colorado and Implications for Iguanodontid-Hadrosaurid Evolution.' *Geological Society of America 38th Annual Meeting, Rocky Mountain Section, Abstracts with Programs* 17: 252.

~ 1985. 'Vanishing Tracks along Alameda

Parkway: Implications for Cretaceous Dinosaurian Paleobiology from the Dakota Group, Colorado.' *Field Guide to Environments of Deposition (and Trace Fossils) of Cretaceous Sandstones of the Western Interior. Society of Economic Paleontologists and Mineralogists, 1985 Midyear Meeting, Golden, Colorado*: 131–42.

~ 1986. *A Guide to Dinosuar Tracksites of the Colorado Plateau and the American Southwest. University of Colorado at Denver Geology Department Magazine, Special Issue* 1: 56 pp.

Lull, R.S. 1911. 'The Reptilia of the Arundel Formation.' *Maryland Geological Survey, Lower Cretaceous*: 173–8.

~ 1911. 'Vertebrata.' *Maryland Geological Survey, Lower Cretaceous*: 183–211.

~ 1921. 'The Cretaceous Armored Dinosaur, *Nodosaurus textilis* Marsh.' *American Journal of Science* 1: 97–126.

Lydekker, R. 1888–1890. *Catalogue of the Fossil Reptilia and Amphibia in the British Museum (Natural History).* Part 1, 309 pp.; Part 2, 307 pp; Part 3, 239 pp.; Part 4, 295 pp. British Museum (Natural History), London.

Mallory, W.W. (ed.). 1972. (See references for Triassic.) (See article by D.P. McGookey, J.D. Haun, L.A. Hale, H.G. Goodell, D.G. McCubbin, R.J. Weimer, and G.R. Wulf.)

McGowan, C. 1972. 'The Systematics of Cretaceous Ichthyosaurs with Particular Reference to the Material from North America.' *University of Wyoming Contributions to Geology* 11: 9–29.

Martin, L.D. 1983. 'The Origin and Early Radiation of Birds.' In A.H. Brush and G.A. Clark Jr., eds, *Perspectives in Ornithology*, pp. 291–338. Cambridge University Press, New York.

Mehl, M.G. 1931. 'Additions to the Vertebrate Record of the Dakota Sandstone.' *American Journal of Science* 21: 441–52.

Miller, H.W., Jr. 1964. 'Cretaceous Dinosaurian Remains from Southern Arizona. *Journal of Paleontology* 38: 378–84.

Moodie, R.L. 1910. 'An Armored Dinosaur from the Cretaceous of Wyoming.' *Kansas Univeristy Science Bulletin* 5: 257–73.

Mossman, D.J., and W.A.S. Sarjeant. 1983. 'The Footprints of Extinct Animals.' *Scientific American* 248: 74–85.

Nelson, M.E., J.H. Madsen, and W.L. Stokes. 1984. 'A New Vertebrate Fauna from the Cedar Mountain Formation (Cretaceous), Emery County, Utah.' Geological Society of America, Abstracts with Program 16: 249.

Norman, D.B. 1980. *On the Ornithischian Dinosaur* Iguanodon bernissartensis *from the Lower Cretaceous of Bernissart (Belgium).*

Institut royal de Sciences naturelles de Belgique Mémoire 178. 103 pp.

~ 1984. 'On the Cranial Morphology and Evolution of Ornithopod Dinosaurs.' *Symposium Zoological Society of London* 52: 521–47.

Ostrom, J.H. 1969. *Osteology of* Deinonychus antirrhopus, *an Unusual Theropod from the Lower Cretaceous of Montana.* Peobody Museum of Natural History Yale University Bulletin 30. 165 pp.

~ 1970. *Stratigraphy and Paleontology of the Cloverly Formation (Lower Cretaceous) of the Bighorn Basin Area, Wyoming and Montana.* Peobody Museum of Natural History Yale University, Bulletin 35. 234 pp.

~ 1976. *On a New Specimen of the Lower Cretaceous Theropod Dinosaur* Deinonychus antirrhopus. Breviora Museum of Comparative Zoology 439. 21 pp.

Patterson, B. 1956. 'Early Cretaceous Mammals and the Evolution of Mammalian Molar Teeth.' *Fieldiana Geology* 13: 1–105.

Perkins, B.F., and C.L. Stewart. 1971. 'Stop 7: Dinosaur Valley State Park.' *Louisiana State University Miscellaneous Publication* 71–1: 56–9.

Pittman, J.G., and D.D. Gillette. 1985. 'A New Sauropod Dinosaur Trackway from the Lower Cretaceous of Arkansas.' Geological Society of America, Abstracts with Programs 17: 187.

Retallack, G., and D.L. Dilcher. 1981. 'A Coastal Hypothesis for the Dispersal and Rise to Dominance of Flowering Plants. In K.J. Niklas (ed.), *Paleobotany, Paleoecology and Evolution*, volume 2, pp. 27–77. Praeger Publishers, New York.

Rubey, W.W. 1973. *New Cretaceous Formations in the Western Wyoming Thrust Sheet.* U.S. Geological Survey Bulletin 1372-I. 35 pp.

Ruiz, A.L. 1985. 'Nota sobre las plumas fósiles del yacimiento eocretácico de "La Pedrera La Cabrúa" en la Sierra del Monjtsec (Prov. Lleida, España).' *Instituto de Estudios Ilerdenses* 1985: 227–38.

Russell, D.A. 1984. *A Check List of the Families and Genera of North American Dinosaurs.* Syllogeus 53. 35 pp.

Sarjeant, W.A.S. 1981. 'In the Footsteps of the Dinosaurs.' *Explorers Journal* 59: 164–71.

Schwab, K.W. 1977. 'Palynology of the Lower Thermopolis and Cloverly Formations (Albian), Gros Ventre Mountains, Wyoming.' *Palynology* 1: 176.

Scott, R.W. 1984. 'Mesozoic biota and Depositional Systems of the Gulf of Mexico–Caribbean Region.' Geological Association of Canada Special Paper 27: 49–64.

Singh, C. 1975. 'Stratigraphic Significance of Early Angiosperm Pollen in the Mid-Cretaceous of Alberta.' Geological Association of Canada Special Paper 13: 365–89.

~ 1983. *Cenomanian Microfloras of the Peace River Area, Northwestern Alberta.* Alberta Research Council Bulletin 44. 322 pp.

Sohn, I.G. 1979. *Nonmarine Ostracodes in the Lakota Formation (Lower Cretaceous) from South Dakota and Wyoming.* U.S. Geological Survey Professional Paper 1069. 24 pp.

Storer, J.E. 1975. 'Dinosaur Tracks, *Columbosauripus ungulatus*, from the Dunvegan Formation (Cenomanian) of Northeastern British Columbia.' *Canadian Journal of Earth Sciences* 12: 1805–7.

Stott, D.F. 1973. *Lower Cretaceous Bullhead Group between Bullmoose Mountain and Tetsa River, Rocky Mountain Foothills, Northeastern British Columbia.* Geological Survey of Canada Bulletin 219. 228 pp.

~ 1983. 'Late Jurassic–Early Cretaceous Foredeeps of Northeastern British Columbia.' *Transactions of the Royal Society of Canada* (IV) 21: 143–53.

~ 1984. 'Cretaceous Sequences of the Foothills of the Canadian Rocky Mountains.' *Canadian Society of Petroleum Geologists Memoir* 9: 85–107.

Stovall, J.W., and W. Langston Jr. 1950. '*Acrocanthosaurus atokensis*, a New Genus and Species of Lower Cretaceous Theropods from Oklahoma.' *American Midland Naturalist* 43: 697–728.

Sues, H.D. 1980. 'Anatomy and Relationships of a New Hypsilophodontid Dinosaur from the Lower Cretaceous of North America.' *Palaeontographica Abt. A* 169: 51–72.

Sues, H.D., and P.M. Galton. 1982. 'The Systematic Position of *Stenopelix valdensis* (Reptilia: Ornithischia) from the Wealden of North-western Germany.' *Palaeontographica Abt. A*, Band 178: 183–90.

Telford, P.G., and H.M. Verma (eds.). 1982. *Mesozoic Geology and Mineral Potential of the Moose River Basin*, Ontario Geological Survey Study 21. 193 pp.

Thayn, G.F., W.D. Tidwell, and W.L. Stokes. 1983. 'Flora of the Lower Cretaceous Cedar Mountain Formation of Utah and Colorado, Part I, *Paraphyllanthoxylon utahense*.' *Great Basin Naturalist* 43: 394–402.

Throckmorton, G.S., J.A. Hopson, and P. Parks. 1981. 'A Redescription of *Toxolophosaurus cloudi* Olson, a Lower Cretaceous Herbivorous Sphenodontid Reptile.' *Journal of Paleontology* 55: 586–97.

Thurmond, J.T. 1971. 'Cartilaginous Fishes of the Trinity Group and Related Rocks (Lower Cretaceous) of North Central Texas.' *Southeastern Geology* 13: 207–27.

Try, C.F., D.G.F. Long, and C.G. Winder. 1984. 'Sedimentology of the Lower Cretaceous Mattagami Formation, Moose River Basin, James Bay Lowlands, Ontario, Canada.' *Canadian Society of Petroleum Geologists Memoir* 9: 345–59.

Tschudy, R.H., B.D. Tschudy, and L.C. Craig. 1984. *Palynological Evaluation of Cedar Mountain and Burro Canyon Formations, Colorado Plateau.* U.S. Geological Survey Professional Paper 1281. 24 pp.

Upchurch, G.R., and J.A. Doyle. 1981. 'Paleoecology of the Conifers *Frenelopsis* and *Pseudofrenelopsis* (Cheirolepidiaceae) from the Cretaceous Potomac Group of Maryland and Virginia.' In R.C. Romans (ed.), *Geobotany* II pp. 167–202. Plenum Press, New York.

Vuke, S.M. 1984. 'Depositional Environments of the Early Cretaceous Western Interior Seaway in Southwestern Montana and the Northern United States.' *Canadian Society of Petroleum Geologists Memoir* 9: 127–44.

Waage, K.M. 1959. 'Stratigraphy of the Inyan Kara Group in the Black Hills.' *U.S. Geological Survey Bulletin* 1081B: 7–89.

~ 1961. *Stratigraphy and Refractory Clayrocks of the Dakota Group along the Northern Front Range, Colorado.* U.S. Geological Survey Bulletin 1102. 154 pp.

Watson, J. 1977. 'Some Lower Cretaceous Conifers of the Cheirolepidiaceae from the U.S.A. and England.' *Palaeontology* 20: 715–49.

Watson, J., and H.L. Fisher. 1984. 'A New Conifer Genus from the Lower Cretaceous Glen Rose Formation, Texas.' *Palaeontology* 27: 719–27.

Weishampel, D.B., and J.B. Weishampel. 1983. 'Annotated Localities of Ornithopod Dinosaurs: Implications to Mesozoic Paleobiogeography.' *Mosasaur* 1: 43–87.

Welles, S.P. 1962. 'A New Species of Elasmosaur from the Aptian of Columbia and a Review of the Cretaceous Plesiosaurs.' *University of California Publications in Geological Science* 44: 1–96.

~ 1970. 'The Longest Neck in the Ocean.' *University of Nebraska News* 50 (9).

Wieland, G.R. 1931. 'Land Types of the Trinity Beds.' *Science* 74: 393–5.

~ 1937. 'Fossil Cycad National Monument.' *Science* 85: 287–9.

Williams, G.D., and C.R. Stelck. 1975. 'Speculations on the Cretaceous Palaeogeography of North America.' Geological Association of Canada Special Paper 13: 1–29.

Williams, G.E., and J.G. Douglas. 1985. 'Comments on Cretaceous Climatic Equability in Polar Regions.' *Palaeogeogeography, Palaeoclimatology, Palaeoecology* 49: 355–9.

Cretaceous Seas

(CHAPTER SIX)

Brower, J.C. 1983. 'The Aerodynamics of *Pterandodon* and *Nyctosaurus*, Two Large Pterosaurs from the Upper Cretaceous of Kansas.' *Journal of Vertebrate Paleontology* 3: 84–124.

Raup, D.M. 1987. 'Mass Extinction: A Commentary.' *Palaeontology* 30: 1–13.

Other references are contained in:

Russell, D.A. (In press). 'Vertebrates in the Cretaceous Interior Sea.' Geological Association of Canada, Special Papers.

~ 1988. 'A Check List of North American Marine Cretaceous Vertebrates, Including Fresh Water Fishes.' Occasional Papers of the Tyrrell Museum of Palaeontology 4. 58 pp.

The Late Cretaceous

(CHAPTERS SEVEN AND EIGHT)

*Archibald, J.D. 1982. *A study of Mammalia and Geology across the Cretaceous-Tertiary Boundary in Garfield County, Montana.* University of California Publications, Geological Sciences 122. 286 pp.

Baird, D. 1986. 'Upper Cretaceous Reptiles from the Severn Formation of Maryland.' *Mosasaur* 3: 63–85.

Baird, D., and J.R. Horner. 1979. 'Cretaceous Dinosaurs of North Carolina.' *Brimleyana* 2: 1–28.

Béland, P., and D.A. Russell. 1978. 'Paleoecology of Dinosaur Provincial Park (Cretaceous), Alberta, Interpreted from the Distribution of Articulated Vertebrate Remains.' *Canadian Journal of Earth Sciences* 15: 1012–24.

Bell, W.A. 1965. *Upper Cretaceous and Paleocene Plants of Western Canada.* Geological Survey of Canada Paper 65–35. 46 pp.

Bird, R.T. 1985. (See references for Preface.)

Boehlke, J.E., and P.L. Abbott. 1986. 'Depositional Environment and Provenance of the Punta Baja Formation (Campanian), El Rosario, Baja California.' Geological Society of America, Abstracts with Program 18 no. 2: 87.

Brown, B. 1913. 'The Skeleton of *Saurolophus*, a Crested Duck-billed Dinosaur from the Edmonton Cretaceous.' *Bulletin of the American Museum of Natural History* 32: 387–93.

~ 1916. '*Corythosaurus casuarius*: Skeleton, Musculature and Epidermis.' *Bulletin of the American Museum of Natural History* 35: 709–16.

~ 1917. 'A Complete Skeleton of the Horned Dinosaur *Monoclonius*, and Description of a Second Skeleton Showing Skin Impressions. *Bulletin of the American Museum of Natural History* 37: 281–306.

Brown, B., and E.M. Schlaikjer. 1942. *The Skeleton of* Leptoceratops with the Description of a New Species. Novitates American Museum of Natural History 1169. 15 pp.

Carpenter, K. 1982. 'Baby Dinosaurs from the Late Cretaceous Lance and Hell Creek Formations and a Description of a New Species of Theropod.' *Contributions to Geology, University of Wyoming* 20: 123–34.

~ 1982. 'The Oldest Late Cretaceous Dinosaurs in North America?' *Mississippi Geology* 3: 11–17.

~ 1982. 'Skeletal and Dermal Armor Reconstruction of *Euoplocephalus tutus* (Ornithischia: Ankylosauridae) from the Late Cretaceous Oldman Formation of Alberta.' *Canadian Journal of Earth sciences* 19: 689–97.

*Carpenter, K., and B. Breithaupt. 1986. 'Latest Cretaceous Occurrence of Nodosaurid Ankylosaurs (Dinosauria, Ornithischia) in Western North America and the Gradual Extinction of the Dinosaurs.' *Journal of Vertebrate Paleontology* 6: 251–7.

Case, T.J. 1978. 'Speculations on the Growth Rate and Reproduction in Dinosaurs.' *Paleobiology* 4: 320–8.

Chamberlain, V.E., and R. St. J. Lambert. 1985. 'Cordilleria, a Newly Defined Canadian Microcontinent.' *Nature* 314: 737–43.

Clemens, W.A., and C.W. Allison. 1985. 'Later Cretaceous Terrestrial Vertebrate Fauna, North Slope, Alaska.' Geological Society of America, Abstracts with Programs, p. 548.

Coe, M.J., D.L. Dilcher, J.O. Farlow, D.M. Jarzen, and D.A. Russell. 1987. (See references to Jurassic.)

Coombs, W.P. 1978. 'Forelimb Muscles of the Ankylosauria (Reptilia, Ornithischia).' *Journal of Paleontology* 52: 642–57.

~ 1978. (See references to Jurassic.)

~ 1979. 'Osteology and Myology of the Hindlimb in the Ankylosauria (Reptilia, Ornithischia).' *Journal of Paleontology* 53: 666–84.

Currie, P.J. 1981. 'Hunting Dinosaurs in Alberta's Great Bonebed.' *Canadian Geographical Journal* 101: 34–9.

~ 1983. (See references to early Cretaceous.)

~ 1987. 'Bird-like Characteristics of the Jaws and Teeth of Troodontid Theropods (Dinosauria: Saurischia).' *Journal of Vertebrate Paleontology* 7: 72–81.

Davies, K.L. 1983. 'Hadrosaurian Dinosaurs of Big Bend National Park, Brewster County, Texas.' MA thesis, University of Texas at Austin. 235 pp.

DeCourten, F.L., and D.A. Russell. 1985. 'A Specimen of *Ornithomimus velox* (Theropoda, Ornithomimidae) from the Terminal Cretaceous Kaiparowits Formation of Southern Utah.' *Journal of Paleontology* 59: 1091–9.

Dodson, P. 1975. 'Taxonomic Implications of Relative Growth in Lambeosaurine Hadrosaurs.' *Systematic Zoology* 24: 37–54.

~ 1983. 'A Faunal Review of the Judith River (Oldman) Formation, Dinosaur Provincial Park, Alberta.' *Mosasaur* 1: 89–118.

~ 1985. *Paleoecology of the Judith River Formation of Southern Alberta.* Report to the National Geographic Society. 22 pp.

Edmund, G. 1985. 'Dinosaur Bones in the Chihuahua Desert.' *Rotunda* 17: 35–9.

Emry, R.J., J.D. Archibald, and C.C. Smith. 1981. 'A Mammalian Molar from the Late Cretaceous of Northern Mississippi.' *Journal of Paleontology* 55: 953–6.

Enos, P. 1983. 'Late Mesozoic Paleogeography of Mexico.' *Rocky Mountain Paleogeography Symposium* 2: 133–57.

Erickson, J.M. 1978. 'Bivalve Mollusk Range Extensions in the Fox Hills Formation (Maestrichtian) of North and South Dakota.' *Annual Proceedings of the North Dakota Academy of Science* 32: 78–89.

Estes, R., and P. Berberian. 1970. *Paleoecology of a Late Cretaceous Vertebrate Community from Montana.* Breviora Museum of Comparative Zoology 343. 35 pp.

Farlow, J.O. 1976. 'A Consideration of the Trophic Dynamics of a Late Cretaceous Large-Dinosaur Community (Oldman Formation).' *Ecology* 57: 841–57.

~ 1980. 'Predator/Prey Biomass Ratios, Community Food Webs and Dinosaur Physiology.' *American Association for the Advancement of Science Selected Symposium* 28: 55–83.

Farlow, J.O., and P. Dodson. 1975. 'The Behavioral Significance of Frill and Horn Morphology in Ceratopsian Dinosaurs.' *Evolution* 29: 353–61.

Fouch, T.D., T.F. Lawton, D.J. Nichols, W.B. Cashion, and W.A. Cobban. 1983. 'Patterns and Timing of Synorogenic Sedimentation in Upper Cretaceous Rocks of Central and Northeast Utah.' *Rocky Mountain Paleogeography Symposium* 2: 305–36.

Fox, R.C. 1976. 'Cretaceous Mammals (*Meniscoessus intermedius*, New Species, and *Alphadon* sp.) from the Lowermost Oldman Formation, Alberta.' *Canadian Journal of Earth Sciences* 13: 1216–22.

~ 1980. 'Upper Cretaceous Terrestrial Vertebrate Stratigraphy of the Gobi Desert

(Mongolian People's Republic) and Western North America.' Geological Association of Canada Special Paper 18: 577–94.

~ 1981. 'Mammals from the Upper Cretaceous Oldman Formation, Alberta v. *Eodelphis* Matthew, and the Evolution of the Stagodontidae (Marsupalia).' *Canadian Journal of Earth Sciences* 18: 350–65.

Fox, R.C., and B.G. Naylor. 1982. 'A Reconsideration of the Relationships of the Fossil Amphibian *Albanerpeton.*' *Canadian Journal of Earth Sciences* 19: 118–28.

Frye, C.I. 1969. *Stratigraphy of the Hell Creek Formation in North Dakota.* North Dakota Geological Survey Bulletin. 54. 65 pp.

Gallagher, W.B. 1986. 'Paleontology, Biostratigraphy, and Depositional Environments of the Cretaceous-Tertiary Transition in the New Jersey Coastal Plain.' *Mosasaur* 3: 1–35.

Galton, P.M., and H.D. Sues. 1983. 'New Data on Pachycephalosaurid Dinosaurs (Reptilia: Ornithischia) from North America.' *Canadian Journal of Earth Sciences* 20: 462–72.

Gastil, R.G. 1986. 'Cretaceous Paleogeography of Peninsular California.' *American Association of Petroleum Geologists Bulletin.* 70: 469.

Gill, J.R., and W.A. Cobban. 1973. *Stratigraphy and Geologic History of the Montana Group and Equivalent Rocks, Montana, Wyoming, and North and South Dakota.* U.S. Geological Survey Professional Paper 776. 37 pp.

Gilmore, C.W. 1946. 'Reptilian Fauna of the North Horn Formation of Central Utah.' U.S. Geological Survey Professional Paper 210c: 25–53.

Gilmore, C.W., and D.R. Stewart. 1945. 'A New Sauropod Dinosaur from the Upper Cretaceous of Missouri.' *Journal of Paleontology* 19: 23–9.

Hamilton, W. 1983. 'Cretaceous and Cenozoic History of the Northern Continents.' *Annals of the Missouri Botanical Garden* 70: 440–58.

Hatcher, J.B., O.C. Marsh, and R.S. Lull. 1907. *The Ceratopsia.* Monographs of the U.S. Geological Survey 49. 330 pp.

Horner, J.R. 1979. 'Upper Cretaceous Dinosaurs from the Bearpaw Shale (Marine) of South-Central Montana with a Checklist of Upper Cretaceous Dinosaur Remains from Marine Sediments in North America.' *Journal of Paleontology* 53: 566–77.

~ 1983. 'Cranial Osteology and Morphology of the Type Specimen of *Maiasaura peeblesorum* (Ornithischia: Hadrosauridae), with Discussion of Its Phylogenetic Position. *Journal of Vertebrate Paleontology* 3: 29–38.

~ 1984. 'The Nesting Behavior of Dinosaurs.'

Scientific American 250: 130–7.

~ 1984. 'Three Ecologically Distinct Vertebrate Faunal Communities from the Late Cretaceous Two Medicine Formation of Montana, with Discussion of Evolutionary Pressures Induced by Interior Seaway Fluctuations.' *Montana Geological Society 1984 Field Conference*: 299–303.

~ 1985. 'Evidence for Polyphylectic Origination of the Hadrosauridae.' *Proceedings of the Pacific Division, American Association for the Advancement of Science* 4: 31–2.

~ 1986. 'Ecologic Interpretations and Morphology of Dinosaur Nests.' *First International Symposium on Dinosaur Tracks and Traces, Albuquerque*: 17.

Hutchinson, P.J., and B.S. Kues. 1985. 'Depositional Environments and Paleontology of Lewis Shale to Lower Kirtland Shale Sequence (Upper Cretaceous), Bisti Area, Northwestern New Mexico.' *New Mexico Bureau of Mines and Mineral Resources Circular* 195: 25–54.

Jacobs, L.L. (ed.). 1980. (See references to Triassic.) (See article by D.A. Russell.)

Jarzen, D.M. 1982. *Palynology of Dinosaur Provincial Park (Campanian) Alberta.* Syllogeus 38. 69 pp.

Jeletzky, J.A. 1971. (See references for early Cretaceous.)

Jenkins, F.A., and D.W. Krause. 1983. 'Adaptations for Climbing in North American Multituberculates (Mammalia).' *Science* 220: 712–15.

Jensen, J.A. 1966. 'Dinosaur Eggs from the Upper Cretaceous North Horn Formation of Central Utah.' *Brigham Young University Geology Studies* 13: 55–67.

Jerzykiewicz, T., and A.R. Sweet. 1986. 'Caliche and Associated Impoverished Palynological Assemblages: An Innovative Line of Paleoclimatic Research onto the Uppermost Cretaceous and Paleocene of Southern Alberta.' Geological Survey of Canada Paper 86–1B: 653–63.

Kaye, J.M., and D.A. Russell. 1973. 'The Oldest Record of Hadrosaurian Dinosaurs in North America.' *Journal of Paleontology* 47: 91–3.

Koster, E.H. 1983. *Field Trip Guidebook, Sedimentology of the Upper Cretaceous Judith River (Belly River) Formation, Dinosaur Provincial Park, Alberta.* Canadian Society of Petroleum Geology, Calgary. 121 pp.

Kurzanov, S.M. 1976. 'Braincase Structure in the Carnosaur *Itemirus* n. gen. and Some Aspects of the Cranial Anatomy of Dinosaurs.' *Paleontological Journal* 1976 (3): 127–37.

Langston, W. 1960. 'The Vertebrate Fauna of the Selma Formation of Alabama, Part 6, The Dinosaurs.' *Fieldiana Geology Memoirs* 3: 313–61.

~ 1981. 'Pterosaurs.' *Scientific American,* February: 122–36.

Lawson, D.A. 1975. 'Pterosaur from the Latest Cretaceous of West Texas: Discovery of the Largest Flying Creature.' *Science* 187: 947–8.

~ 1976. '*Tyrannosaurus* and *Torosaurus,* Maestrichtian Dinosaurs from Trans-Pecos, Texas.' *Journal of Paleontology* 50: 158–64.

Lehman, T.M. 1982. 'A Ceratopsian Bone Bed from the Aguja Formation (Upper Cretaceous), Big Bend National Park, Texas.' MA thesis, University of Texas at Austin. 210 pp.

~ 1985. 'Depositional Environments of the Naashoibito Member of the Kirtland Shale, Upper Cretaceous, San Juan Basin, New Mexico.' *New Mexico Bureau of Mines and Mineral Resources Circular* 195: 55–79.

Lerbekmo, J.F. 1985. 'Magnetostratigraphic and Biostratigraphic Correlations of Maastrichtian to Early Paleocene Strata between South-Central Alberta and Southwestern Saskatchewan.' *Canadian Society of Petroleum Geologists* 33: 213–26.

Lerbekmo, J.F., and K.C. Coulter. 1985. 'Late Cretaceous to Early Tertiary Magnetostratigraphy of a Continental Sequence: Red Deer Valley, Alberta.' *Canadian Journal of Earth Sciences* 22: 567–83.

Lillegraven, J.A. 1972. *Preliminary Report on Late Cretaceous Mammals from the El Gallo Formation, Baja California del Norte, Mexico.* Los Angeles County Museum Contributions in Science 232. 11 pp.

~ 1976. 'A New Genus of Therian Mammal from the Late Cretaceous "El Gallo Formation," Baja California, Mexico.' *Journal of Paleontology* 50: 437–43.

Lockley, M. 1986. (See references for Early Cretaceous.)

Lorenz, J.C., and W. Gavin. 1984. 'Geology of the Two Medicine Formation and the Sedimentology of a Dinosaur Nesting Ground.' *Montana Geological Society 1984 Field Conference*: 175–86.

Lucas, S.G., N.J. Mateer, A.P. Hunt, and F.M. O'Neill. 1987. 'Dinosaurs, the Age of the Fruitland and Kirtland Formations, and the Cretaceous-Tertiary Boundary in the San Juan Basin, New Mexico.' Geological Society of America Special Paper 209: 35–50.

Lucas, S.G., J.K. Rigby and B.S. Kues (eds). 1981. *Advances in San Juan Basin Paleontology.* University of New Mexico Press, Albuquerque. 393 pp. (See articles by T.M. Lehman, S.G. Lucas, and W.D. Tidwell; S.R. Ash and L.R. Parker.)

Lull, R.S. 1915. 'The Mammals and Horned Dinosaurs of the Lance Formation of Niobrara County, Wyoming.' *American Journal of Science* 40: 319–48.

~ 1933. *A Revision of the Ceratopsia or Horned Dinosaurs.* Memoirs of the Peabody Museum of Yale University 3. 175 pp.

Lull, R.S., and N.E. Wright. 1942. *Hadrosaurian Dinosaurs of North America.* Geological Society of America Special Papers 40. 242 pp.

Lupton, C.D., D. Gabriel, and R.M. West. 1980. 'Paleobiology and Depositional Setting of a Late Cretaceous Vertebrate Locality, Hell Creek Formation, McCone County, Montana.' *Contributions to Geology, University of Wyoming* 18: 117–26.

Mallory, W.W. (ed.). 1972. (See references for Jurassic.) (See articles by D.P. McGookey, J.D. Haun, L.A. Hale, H.G. Goodell, D.G. McCubbin, R.J. Weimer, and G.R. Wulf.)

Marshall, L.G., C. de Muizon, M. Gayet, A. Lavenu, and B. Sige. 1985. 'The "Rosetta Stone" for Mammalian Evolution in South America.' *National Geographic Research* 1: 274–88.

Martin, L.D. 1983. 'Origin and Early Radiation of Birds.' In A.H. Brush and G.A. Clark Jr., eds, *Perspectives in Ornithology,* pp. 291–338. Cambridge University Press, New York.

Maryanska, T. 1977. 'Ankylosauridae (Dinosauria) from Mongolia.' *Acta Palaeontologia Polonica* 37: 85–151.

Molenaar, C.M. 1983. 'Major Depositional Cycles and Regional Correlations of Upper Cretaceous Rocks, Southern Colorado Plateau and Adjacent Areas.' *Rocky Mountain Paleogeography Symposium* 2: 201–24.

Molnar, R.E. 1974. 'A Distinctive Theropod Dinosaur from the Upper Cretaceous of Baja California (Mexico).' *Journal of Paleontology* 48: 1009–17.

~ 1978. 'A New Theropod Dinosaur from the Upper Cretaceous of Central Montana.' *Journal of Paleontology* 52: 73–82.

~ 1980. 'An Albertosaur from the Hell Creek Formation of Montana.' *Journal of Paleontology* 54: 102–8.

Morris, W.J. 1967. 'Baja California: Late Cretaceous Dinosaurs.' *Science* 155: 1539–41.

~ 1972. 'A Giant Hadrosaurian Dinosaur from Baja California.' *Journal of Paleontology* 46: 777–9.

~ 1973. 'Mesozoic and Tertiary Vertebrates in Baja California.' *National Geographic Research Reports,* 1966: 197–209.

~ 1973. 'A Review of Pacific Coast Hadrosaurs.' *Journal of Paleontology* 47: 551–61.

~ 1981. 'A New Species of Hadrosaurian Dinosaur from the Upper Cretaceous of Baja California – ? *Lambeosaurus laticaudus.*' *Journal of Paleontology* 55:

453–62.

Naylor, B.G. (ed.). 1986. *Dinosaur Systematics Symposium, Tyrrell Museum of Palaeontology, Field Trip Guidebook to Dinosaur Provincial Park.* Tyrrell Museum of Palaeontology, Drumheller, Alberta. 56 pp.

Nesov, L.A. 1981. 'Late Cretaceous Flying Lizards from the Kyzyl-Kumy Region.' *Paleontological Journal* 1981 (4): 98–104.

~ 1984. 'Upper Cretaceous Pterosaurs and Birds from Central Asia.' *Paleontological Journal* 1984 (1): 47–57.

Nicholls, E.L. 1972. 'Fossil Turtles from the Campanian Stage of Western North America.' MA thesis, University of Calgary. 116 pp.

Nicholls, E.L., and A.P. Russell. 1985. 'Structure and Function of the Pectoral Girdle and Forelimb of *Struthiomimus altus* (Theropoda: Ornithomimidae).' *Palaeontology* 28: 643–77.

Norris, G., P. Dobell, and J. Gittins. 1980. 'Late Cretaceous *Normapolles* in Karstic Carbonatite Terraine, Southern Moose River Basin, Ontario, Canada.' Abstracts Fifth International Palynological Conference, p. 289.

Oriel, S.S., and J.I. Tracey Jr. 1970. *Uppermost Cretaceous and Tertiary Stratigraphy of Fossil Basin, Southwestern Wyoming.* U.S. Geological Survey Professional Paper 635. 53 pp.

Osborn, H.F. 1917. (See references for Jurassic.)

Ostrom, J.H. 1978. '*Leptoceratops gracilis* from the "Lance" Formation of Wyoming.' *Journal of Paleontology* 52: 697–704.

*Ostrom, J.H., and P. Wellnhofer. 1986. 'The Munich Specimen of *Triceratops* with a Revision of the Genus.' *Zitteliana* 14: 111–58.

Padian, K. 1984. 'A Large Pterodactyloid Pterosaur from the Two Medicine Formation (Campanian) of Montana.' *Journal of Vertebrate Paleontology* 4: 516–24.

Page, V.M. 1981. 'Dicotyledonous Wood from the Upper Cretaceous of Central California. III Conclusions.' *Journal of the Arnold Arboretum* 62: 437–55.

Parker, L.R., and J.K. Balsley. 1986. 'Dinosaur Footprints in Coal Mine Roof Surfaces from the Cretaceous of Utah.' First International Symposium on Dinosaur Tracks and Traces, Albuquerque, Abstracts with Program, p. 22.

Parker, S.E., and R.W. Jones. 1985. 'Upper Cretaceous and Lower Tertiary Rocks of the Bighorn Basin, Wyoming: A New Look at the Basin's Stratigraphy and Tectonic History.' Geological Society of America Abstracts with Program 17(7): 685.

Rahn, H., C.V. Paganelli, and A. Ar. 1975.

'Relation of Avian Egg Weight to Body Weight.' *Auk* 92: 750–65.

Reif, W.E., and F. Westphal. 1984. (See references for Triassic.) (See articles by J.F. Bonaparte, P.J. Currie and P. Dodson.)

Rozhdestvenskii, A.K. 1964. 'New Data on Occurrences of Dinosaurs in Kazakhstan and Central Asia.' *Tashkent State University, Scientific Publications, Geology* 234: 227–41.

~ 1968. 'Hadrosaurs of Kazakhstan.' In V.S. Vanin (ed.), *Upper Paleozoic and Mesozoic Amphibians and Reptiles of the* U.S.S.R., pp. 97–141. Nauka Publications, Moscow.

~ 1977. 'Kansai – a Cretaceous Vertebrate Site in Fergana.' *Yearbook of the All-Union Paleontological Society* 20: 235–47.

Russell, D.A. 1970. 'A Skeletal Reconstruction of *Leptoceratops gracilis* from the Upper Edmonton Formation (Cretaceous) of Alberta.' *Canadian Journal of Earth Sciences* 7: 181–4.

~ 1970. *Tyrannosaurs from the Late Cretaceous of Western Canada.* Publications in Palaeontology, National Museum of Natural Sciences, National Museums of Canada 1. 34 pp.

~ 1972. 'Ostrich Dinosaurs from the Late Cretaceous of Western Canada.' *Canadian Journal of Earth Sciences* 9: 375–402.

~ 1977. *A Vanished World: The Dinosaurs of Western Canada.* National Museum of Natural Sciences, National Museums of Canada. 142 pp.

~ 1983. 'A Canadian Dinosaur Park.' *Terra* 21: 3–9.

~ 1984. (See references for early Cretaceous.)

Russell, L.S. 1935. 'Fauna of the Upper Milk River Beds, Southern Alberta.' *Transactions of the Royal Society of Canada* 29: section IV 117–28.

~ 1964. *Cretaceous Non-marine Faunas of Northwestern North America.* Royal Ontario Museum Life Sciences Contribution 61. 24 pp.

Schwimmer, D.R. 1986. 'Late Cretaceous Fossils from the Blufftown Formation (Campanian) in Western Georgia.' *Mosasaur* 3: 109–23.

Seymour, R.S. 1979. 'Dinosaur Eggs: Gas Conductance through the Shell, Water Loss during Incubation and Clutch Size.' *Paleobiology* 5: 1–11.

Shilin, P.V. and Y.V. Suslov. 1982. 'A Hadrosaur from the Northeastern Aral Region.' *Paleontological Journal* 1982 (1): 131–5.

Shoemaker, E.M., M. Steiner, J.E. Fassett, and R.H. Tschudy. 1984. 'Magnetostratigraphy of Upper Cretaceous Rocks at Mesa Portales, New Mexico.' Geological Society of America, Abstracts with Programs 16 (4): 255.

Sigleo, W.R., and J. Reinhardt. 1985.

'Paleosols in Some Cretaceous Environments of the Southeastern United States.' Geological Society of America, Abstracts with Programs 1985, p. 716.

Sloan, R.E. 1964. *The Cretaceous System in Minnesota.* Minnesota Geological Survey Report of Investigations 5. 64 pp.

Spicer, R.A. 1986. ''Late Cretaceous North Polar Vegetation and Climate.' Abstracts, Annual Conference Palaeontological Association, Leicester 1986, p. 21.

Spieker, E.M. 1960. 'The Cretaceous-Tertiary Boundary in Utah.' *International Geological Congress, Norden* 5: 14–24.

Tait, J., and B. Brown. 1928. 'How the Ceratopsia Carried and Used Their Head.' *Transactions of the Royal Society of Canada* 22: section v: 13–23.

Tarduno, J.A., and W. Alvarez. 1985. 'Paleolatitudes of Franciscan Limestones.' *Geology* 13: 741.

Tschudy, R.H. 1965. 'An Upper Cretaceous Deposit in the Appalachian Mountains.' U.S. Geological Survey Professional Paper 525B: 64–8.

~ 1981. 'Geographic Distribution and Dispersal of Normapolles Genera in North America.' *Review of Palaeobotany and Palynology* 35: 283–314.

Wall, W.P. and P.M. Galton. 1979. 'Notes on Pachycephalosaurid Dinosaurs (Reptilia: Ornithischia) from North America, with Comments on Their Status as Ornithopods.' *Canadian Journal of Earth Sciences* 16: 1176–86.

Weishampel, D.B., and J.A. Jensen. 1979. '*Parasaurolophus* (Reptilia: Hadrosauridae) from Utah.' *Journal of Paleontology* 53: 1422–7.

Williams, D.L.G., R.S. Seymour, and P. Kerourio. 1984. 'Structure of Fossil Dinosaur Eggshell from the Aix Basin, France.' *Palaeogeography, Palaeoclimatology, Palaeoecology* 45: 23–37.

Williams, G.D., and C.F. Burk. 1966. 'Upper Cretaceous.' In R.G. McCrossan and R.P. Glaister (eds.), *Geological History of Western Canada,* pp. 169–89. Alberta Association of Petroleum Geologists, Calgary.

Wolfe, J.A., J.A. Upchurch, and R. Garland Jr. 1987. North American Nonmarine Climates and Vegetation during the Late Cretaceous.' *Paleogeography, Paleoclimatology, Paleoecology* 61: 33–77.

Yefimov, M.B. 1975. 'Late Cretaceous Crocodiles of Soviet Central Asia and Kazakhstan.' *Paleontological Journal* 3: 146–9.

~ 1982. 'A Two-Fanged Crocodile from the Upper Cretaceous in Tadzhikistan.' *Paleontological Journal* 4: 103–4.

Yorath, C.J., and D.G. Cook. 1984. 'Mesozoic and Cenozoic Depositional History of the Northern Interior Plains of Canada.'

Canadian Society of Petroleum Geologists Memoir 9: 69–83.

Extinction of the Dinosaurs

(CHAPTER NINE)

Alvarez, L., W. Alvarez, F. Asaro, and H.V. Michel. 1980. 'Extraterrestrial Cause for the Cretaceous-Teritiary Extinction.' *Science* 208: 1095–1108.

Alvarez, W. 1986. 'Toward a Theory of Impact Crises.' *Eos* 67: 649–58.

Arthur, M.A., J.C. Zachos, and D.S. Jones. 1987. 'Primary Productivity and the Cretaceous/Tertiary Boundary Event in the Oceans.' *Cretaceous Research* 8: 43–54.

Case, J.A., and M.O. Woodburne. 1986. 'South American Marsupials: A Successful Crossing of the Cretaceous-Tertiary Boundary.' *Palaios* 1: 413–16.

*Clemens, W.A. 1986. 'Evolution of the Terrestrial Vertebrate Fauna during the Cretaceous-Tertiary Transition.' In D.K. Elliott (ed.), *Dynamics of Extinction,* pp. 63–85. John Wiley and Sons, New York.

*Fassett, J.E., S.G. Lucas, and F.M. O'Neill. 1987. 'Dinosaurs, Pollen and Spores, and the Age of the Ojo Alamo Sandstone, San Juan Basin, New Mexico.' Geological Society of America, Special Paper 209: 17–34.

Hartnady, C.J. 1986. 'Amirante Basin, Western Indian Ocean: Possible Impact Site of the Cretaceous/Tertiary Extinction Bolide?' *Geology* 14: 423–6.

*Hutchinson, J.H., and J.D. Archibald. 1986. 'Diversity of Turtles across the Cretaceous/Tertiary Boundary in Northeastern Montana.' *Palaeogeography, Palaeoclimatology, Palaeoecology* 55: 1–22.

Jablonski, D. 1986. 'Background and Mass Extinctions: The Alternation of Macroevolutionary Regimes.' *Science* 231: 129–33.

Kitchell, J.A., D.L. Clark, and N.M. Gombos. 1986. 'Biological Selectivity of Extinction: A Link between Background and Mass Extinction.' *Palaios* 1: 504–11.

Leahy, G.D., M.D. Spoon, and G.J. Retallack. 1985. 'Linking Impacts and Plant Extinctions.' *Nature* 318: 318.

McNaughton, S.J., R.W. Ruess, and M.B. Coughenour. 1986. 'Ecological Consequences of Nuclear War.' *Nature* 321: 483–7.

Nichols, D.J., D.M. Jarzen, C.J. Orth, and P.Q. Oliver. 1986. 'Palynological and Iridium Anomalies at Cretaceous-Tertiary Boundary, South-Central Saskatchewan.' *Science* 231: 714–17.

Pollack, J.B., O.B. Toon, T.P. Ackerman, C.P. McKay, and R.P. Turco. 1983.

'Environmental Effects of an Impact-Generated Dust Cloud: Implications for the Cretaceous-Tertiary Extinctions.' *Science* 219: 287–9.

Preisinger, A., E. Zobetz, A.J. Gratz, R. Lahodynsky, M. Becke, H.J. Mauritsch, G. Eder, F. Grass, R. Rogl, H. Stradner, and R. Surenian. 1986. 'The Cretaceous/Tertiary Boundary in the Gosau Basin, Austria.' *Nature* 322: 794–9.

Raup, D.M., and J.J. Sepkoski Jr. 1986. 'Periodic Extinction of Families and Genera.' *Science* 231: 833–6.

Russell, D.A. 1984. 'The Gradual Decline of the Dinosaurs – Fact or Fallacy?' *Nature* 307: 360–1.

Saito, T., T. Yamanoi, and K. Kaiho. 1986. 'End-Cretaceous Devastation of Terrestrial Flora in the Boreal Far East.' *Nature* 323: 253–5.

Seitz, R. 1986. 'Siberian Fire as ''Nuclear Winter'' Guide.' *Nature* 323: 116–17.

*Sloan, R.E., J. Keith Rigby Jr., L.M. Van Valen, and D. Gabriel. 1986. 'Gradual Dinosaur Extinction and Simultaneous Ungulate Radiation in the Hell Creek Formation.' *Science* 232: 629–33; see also 234: 1170–5.

Smit, J., and A.J.T. Romein. 1985 'A Sequence of Events across the Cretaceous-Tertiary Boundary.' *Earth and Planetary Science Letters* 74: 155–70.

Thierstein, H.R. 1982. 'Terminal Cretaceous Plankton Extinctions: A Critical Assessment.' Geological Society of America Special Paper 190: 385–99.

Tschudy, R.H., C.L. Pillmore, C.J. Orth, J.S. Gilmore, and J.D. Knight. 1984. 'Disruption of the Terrestrial Plant Ecosystem at the Cretaceous-Tertiary Boundary, Western Interior.' *Science* 225: 1030–2.

Tschudy, R.H., and B.D. Tschudy. 1986. 'Extinction and Survival of Plant Life Following the Cretaceous/Tertiary Boundary Event, Western Interior, North America.' *Geology* 14: 667–70.

*Van Valen, L.M. 1984. 'Catastrophes, Expectations, and the Evidence.' *Paleobiology* 10: 121–37.

Wolbach, W.S., R.S. Lewis, and E. Anders. 1985. 'Cretaceous Extinctions: Evidence for Wildfires and Search for Meteoritic Material.' *Science* 230: 167–70; see also 234: 261–4.

Wolfe, J.A., and G.R. Upchurch Jr. 1986. 'Vegetation, Climatic and Floral Changes at the Cretaceous-Tertiary Boundary.' *Nature* 324: 148–52.

Meaning of the Dinosaurs

(CHAPTER TEN)

Bauchot, R. 1965. 'La placentation chez les reptiles.' *L'Année biologique* 4: 547–75.

Benton, M.J. 'The Red Queen Put to the Test.' *Nature* 313: 734–5.

Billingham, J. (ed.). 1981. *Life in the Universe*. National Aeronautics and Space Administration, Conference Publication 2156. 451 pp.

Cailleux, A. 1971. 'Le temps et les échelons de l'évolution.' In J. Zeman (ed.), *Time and Science in Philosophy*, pp. 135–45. Elsevier, Amsterdam.

~ 1985. 'Ressemblances et différences entre les vivants et l'inanime.' *110e Congrès national des Sociétés savantes, Montpellier, sciences, fasc.* 1: 63–71.

Chaloner, W.G., and A. Sheerin. 1979. 'Devonian Macrofloras.' *Special Papers in Palaentology* 23: 145–61.

*Cherfas, J. 1984. 'The Difficulties of Darwinism.' *New Scientist* 1410: 28–30. (Summary of a presentation by S. J. Gould).

Corwin, M. 1983. 'From Chaos to Consciousness.' *Astronomy* 1983: 15–22.

Couper, H. 1986. 'In Search of Solar Systems.' *New Scientist*, 13 November: 34–9.

Flessa, K.W., and D. Jablonski. 1985. 'Declining Phanerozic Background Extinction Rates: Effect of Taxonomic Structure?' *Nature* 313: 216–18.

Ghiselin, M.T. 1984. '*Peripatus* as a Living Fossil.' In N. Eldredge and S. M. Stanley (eds.), *Living Fossils*, pp. 214–17. Springer Verlag, New York.

Gingerich, P.D. 1984. 'Mammalian Diversity and Structure.' *University of Tennessee Studies in Geology* 8: 1–19.

Ho, Mae-wan, P. Saunders, and S. Fox. 1986. 'A New Paradigm for Evolution.' *New Scientist*, 27 February: 41–3.

Jablonski, D. 1986. (See references to the extinction of the dinosaurs).

Jerison, H.J. 1973. *Evolution of the Brain and Intelligence*. Academic Press, New York. 420 pp.

Johnston, R.F., and R.K. Selander. 1964.

'House Sparrows: Rapid Evolution of Races in North America.' *Science* 144: 548–50.

Jones, J.S. 1985. 'The Point of a Toucan's Bill.' *Nature* 315: 182–3.

Knoll, A.H. 1985. 'Patterns of Evolution in the Archean and Proterozoic Eons.' *Paleobiology* 11: 53–64.

Lovelock, J.E., and M. Whitfield. 1982. 'Life Span of the Biosphere.' *Nature* 296: 561–3.

Milne, D., D. Raup, J. Billingham, K. Niklaus, and K. Padian (eds.). 1985. *The Evolution of Complex and Higher Organisms*. National Aeronautics and Space Administration, SP-478. 193 pp.

Morrison, P., J. Billingham, and J. Wolfe (eds.). 1977. *The Search for Extraterrestrial Life*. National Aeronautics and Space Administration, SP-419. 276 pp.

Niklas, K.J., B.H. Tiffney, and A.H. Knoll. 1983. 'Patterns in Vascular Land Plant Diversification.' *Nature* 303: 614–16.

O'Brien, S.J., W.G. Nash, D.E. Wildt, M.E. Bush, and R.E. Benveniste. 1985. 'A Molecular Solution to the Riddle of the Giant Panda's Phylogeny.' *Nature* 317: 140–4.

Packard, A. 1972. 'Cephalopods and Fish: The Limits of Convergence.' *Biological Reviews* 47: 241–307.

Papagianis, M.D. 1985. 'Commission 51: Search for Extraterrestrial Life.' *International Astronomical Union Transactions* 19A: 713–23.

Pettersson, M. 1978. 'Acceleration in Evolution, before Human Times.' *Journal of Social Biological Structures* 1: 201–6.

Raup, D.M. 1985. 'ETI without Intelligence.' In E. Regis Jr, ed, *Extraterrestrials, Science and Alien Intelligence*, pp. 31–42. University Press, Cambridge.

Riedl, R. 1978. *Order in Living Organisms*. John Wiley and Sons, New York. 313 pp.

Robinson, M.H. 1979. 'Untangling Tropical Biology.' *New Scientist*, 3 May: 378–81.

Rodman, P.S., and H.M. McHenry. 1980. 'Bioenergetics and the Origin of Hominid Bipedalism.' *American Journal of Physical Anthropology* 52: 103–6.

Russell, D.A. 1981. 'Speculations on the Evolution of Intelligence.' In John Billingham (ed.), *Life in the Universe*, pp.

259–75. MIT Press, Cambridge, Massachusetts.

~ 1983. 'Exponential Evolution: Implications for Intelligent Extraterrestrial Life.' *Advances in Space Research* 3: 95–103.

~ 1987. (See references for Preface.)

Russell, D.A. and R. Sequin. 1982. *Reconstructions of the Small Cretaceous Theropod* Stenonychosaurus inequalis *and a Hypothetical Dinosauroid*. Syllogeus 37. 43 pp.

Sagan, C. 1977. *The Dragons of Eden: Speculations on the Evolution of Human Intelligence*. Random House, New York. 263 pp.

Scott, A. 1985. 'Update on Genesis.' *New Scientist*, 2 May: 30–3.

Sepkoski, J.J. 1979. 'A Kinetic Model of Phanerozoic Taxonomic Diversity. I. Analysis of Marine Orders.' *Paleobiology* 4: 223–51.

~ 1979. 'A Kinetic Model of Phanerozoic Taxonomic Diversity. II. Early Phanerozoic Families and Multiple Equilibria.' *Paleobiology* 5: 222–51.

Sibley, C.G., and J.E. Ahlquist. 1986. 'Reconstructing Bird Phylogeny by Comparing DNA's.' *Scientific American* 254 (2): 82–92.

*Simpson, G.G. 1964. 'The Nonprevalence of Humanoids.' *Science* 143: 769–75.

Slijper, E.J. 1942. 'Biologic-Anatomical Investigations on the Bipedal Gait and Upright Posture in Mammals, with Special Reference to a Little Goat, Born without Forelimbs.' *Nederlandse Akademie van Wetenschapen, Proceedings* 45: 288–95, 407–15.

Tucker, V.A. 1975. 'The Energetic Cost of Moving About.' *Scientific American* 63: 413–19.

Van Valen, L. 1985. 'A Theory of Origination and Extinction.' *Evolutionary Theory* 7: 133–42.

Van Valkenburgh, B. 1985. 'Locomotor Diversity within Past and Present Guilds of Large Predatory Mammals.' *Paleobiology* 11: 406–28.

Wyles, J.S., J.G. Kunkel, and A.C. Wilson. 1983. 'Birds, Behavior and Anatomical Evolution.' *Proceedings National Academy of Sciences* U.S.A. 80: 4394–7.

Classification
of
Dinosaurs

The following scheme is a compromise between a traditional classification and modern phylogenetic classifications of saurischian (Gauthier 1986) and ornithischian (Soreno 1986) dinosaurs. Only dinosaurian names cited in the text are listed below, and the place of citation can be found by consulting the index.

Gauthier, J. 1986. 'Saurischian Monophyly and the Origin of Birds.' *Memoirs of the California Academy of Sciences* 8:1–55
Parris, D.C., B.S. Grandstaff, B.L. Stinchcomb, and R. Denton 1988. 'Chronister, the Missouri Dinosaur Site.' *Journal of Vertebrate Paleontology* 8 (Abstracts):23A
Soreno, D. 1986. 'Phyogeny of the Bird-Hipped Dinosaurs (Order Ornithischia).' *National Geographic Research* 2:234–56

Superorder Dinosauria
Ancestral dinosaurs
(neither saurischian nor ornithischian)
   Family Staurikosauridae
    Staurikosaurus

ORDER SAURISCHIA
(lizard-hipped dinosaurs)
Ancestral saurischian, not assigned to a suborder
   Family Herrososauridae
    Herrerosaurus

SUBORDER THEROPODA
Infraorder Ceratosauria
   Family Ceratosauridae
    Ceratosaurus
    Coelophysis
    Dilophosaurus
    Podokesaurus
    Segisaurus
    Syntarsus

   Family Baryonychidae
    Baryonyx

   Family Abelisauridae
    Carnotaurus

   Family Noasauridae (noasaurs)

Infraorder not assigned, small theropods of uncertain relationships:
    Aristosuchus
    Coelurus
    Compsognathus
    Longosaurus
    Marshosaurus
    Ornitholestes
    Stokesosaurus

Infraorder Carnosauria
   Family Megalosauridae
   (possibly an artificial group)
    Acrocanthosaurus
    Altispinax

    Megalosaurus
    Torvosaurus

   Family Allosauridae
    Allosaurus

   Family Dryptosauridae
    Dryptosaurus

   Family Tyrannosauridae (tyrannosaurs)
    Albertosaurus
    Daspletosaurus
    Tyrannosaurus

Infraorder Ornithomimosauria
   Family Ornithomimidae (ostrich dinosaurs)
    Archaeornithomimus
    Dromiceiomimus
    Elaphrosaurus
    Ornithomimus
    Struthiomimus

Infraorder Deinonychosauria
   Family Dromaeosauridae
    Deinonychus
    Dromaeosaurus
    Saurornitholestes

   Family Troodontidae
    Troodon (formerly *Stenonychosaurus*)

Infraorder Oviraptorosauria (oviraptors)
   Family Caenagnathidae
    Cheirostenotes
    Microvenator

SUBORDER SAUROPODOMORPHA
Infraorder Prosauropoda (prosauropods)
   Family Anchisauridae (anchisaurs)
    Anchisaurus
    Massospondylus
    Plateosaurus

   Family Melanorosauridae
   Family Segnosauridae (segnosaurs)

Infraorder Sauropoda
(sauropods, brontosaurs)
   Family Vulcanodontidae
    Vulcanodan

   Family Cetiosauridae (cetiosaurs)
    Haplocanthosaurus
    Shunosaurus

   Family Euhelopodidae
    Mamenchisaurus

   Family Diplodocidae
   (whip-tailed sauropods)
    Apatosaurus (formerly *Brontosaurus*)
    Barosaurus
    Diplodocus (also *Supersaurus?*)

   Family Camarasauridae
   (spoon-toothed sauropods)
    Camarasaurus

   Family Brachiosauridae (giraffe dinosaurs)
    Brachiosaurus (also *Ultrasaurus?*)
    Pelorosaurus
    Pleurocoelus

   Family Titanosauridae
    Alamosaurus
    Titanosaurus

ORDER ORNITHISCHIA
(bird-hipped dinosaurs)
Ancestral ornithischians, not assigned to a suborder
   Family Fabrosauridae
    Lesothosaurus
    Nanosaurus
    Othnelia
    Technosaurus

SUBORDER ORNITHOPODA
Infraorder Hypsilophodontia
   Family Heterodontosauridae (heterodontosaurs)
    Pisanosaurus (reference to heterodontosaurs is uncertain)

   Family Hypsilophodontidae (hypsilophodonts)
    Dryosaurus
    Hypsilophodon
    Tenontosaurus
    Valdosaurus
    Yandusaurus
    Zephyrosaurus

   Family Thescelosauridae
    Thescelosaurus

Infraorder Iguanodontia
   Family Iguanodontidae (iguanodonts)
    Camptosaurus
    Iguanodon

   Family Hadrosauridae
   (hadrosaurs, duck-billed dinosaurs)

   ~ Subfamily Hadrosaurinae
   (flatheaded hadrosaurs)
    Aralosaurus
    Brachylophosaurus
    Claosaurus
    Edmontosaurus
    Hadrosaurus
    Hypsibema (see Parris et al. 1988, references under 'classification of dinosaurs')
    Kritosaurus
    Lophorothon
    Maiasaura
    Prosaurolophus
    Saurolophus
    Shatungosaurus

   ~ Subfamily Lambeosaurinae
   (lambeosaurs, tubeheaded hadrosaurs)
    Barsboldia
    Corythosaurus

Hypacrosaurus
Jaxartosaurus
Lambeosaurus
Parasaurolophus

Ancestral armoured dinosaurs, not assigned to suborder
Scelidosaurus
Scutellosaurus

SUBORDER STEGOSAURIA
Family Stegosauridae (plated dinosaurs)
Huayangosaurus
Lexovisaurus
Stegosaurus

SUBORDER ANKYLOSAURIA
(armoured dinosaurs)
Family Nodosauridae
Edmontonia
Hierosaurus
Hoplitosaurus
Hylaeosaurus

Nodosaurus
Panoplosaurus
Polacanthus
Priconodon
Sauropelta
Silvisaurus

Family Ankylosauridae
Ankylosaurus
Euoplocephalus

SUBORDER PACHYCEPHALOSAURIA
Family Homalocephalidae
Yaverlandia

Family Pachycephalosauridae
(dome-headed dinosaurs)
Gravitholus
Ornatotholus
Pachycephalosaurus
Stegoceras
Stygimoloch

SUBORDER CERATOPSIA
Family Psittacosauridae (psittacosaurs)

Family Protoceratopsidae (hornless 'horned' dinosaurs)
Leptoceratops

Family Ceratopsidae (horned dinosaurs)

~ Subfamily Centrosaurinae
(centrosaurines)
Centrosaurus
Monoclonius
Pachyrhinosaurus
Styracosaurus

~ Subfamily Chasmosaurinae
(chasmosaurines)
Anchiceratops
Arrhinoceratops
Chasmosaurus
Diceratops
Eoceratops
Pentaceratops
Torosaurus
Triceratops

Index

The type selected for the text is Meridien,
designed by A. Frutiger in 1957, and
for the display, ITC Veljovic Medium, designed
by J. Veljovic in 1984.
The type was set by Q Composition Inc., Canada.
The paper used is OJI White-A, and the book
was printed and bound by Book Art Inc.

This book was designed by F. Newfeld.